Mirko Wächter

Learning and Execution of Object Manipulation Tasks on Humanoid Robots

Karlsruhe Series on Humanoid Robotics

Edited by Prof. Dr.-Ing. Tamim Asfour

Vol. 03

Learning and Execution of Object Manipulation Tasks on Humanoid Robots

by
Mirko Wächter

KIT Scientific Publishing

Dissertation, Karlsruher Institut für Technologie
KIT-Fakultät für Informatik

Tag der mündlichen Prüfung: 8. Mai 2017
Referenten: Prof. Dr.-Ing. Tamim Asfour, Prof. Dr. Florentin Wörgötter

Impressum

KIT Scientific Publishing

Karlsruher Institut für Technologie (KIT)
KIT Scientific Publishing
Straße am Forum 2
D-76131 Karlsruhe

KIT Scientific Publishing is a registered trademark
of Karlsruhe Institute of Technology.
Reprint using the book cover is not allowed.

www.ksp.kit.edu

Print on Demand 2018 – Gedruckt auf FSC-zertifiziertem Papier

ISSN 2512-0875
ISBN 978-3-7315-0749-9
DOI 10.5445/KSP/1000078313

Learning and Execution of Object Manipulation Tasks on Humanoid Robots

Zur Erlangung des akademischen Grades eines

Doktors der Ingenieurwissenschaften

der KIT-Fakultät für Informatik
des Karlsruher Instituts für Technologie (KIT)

genehmigte

Dissertation

von

Dipl.-Inform. Mirko Wächter

aus Melle

Tag der mündlichen Prüfung:	8. Mai 2017
Referent:	Prof. Dr.-Ing. Tamim Asfour
Korreferent:	Prof. Dr. Florentin Wörgötter

Deutsche Zusammenfassung

Roboter hielten in den letzten Jahren mehr und mehr Einzug in verschiedene Bereiche unseres Lebens. In Fertigungsstraßen in der Industrie sind sie schon länger ein fester Bestandteil, aber auch im alltäglichen Leben werden zunehmend mehr Roboter eingesetzt. Staubsaugerroboter reinigen unsere Wohnungen, Plüschroboter werden zur Therapie und Unterhaltung eingesetzt und in manchen Bars bereiten sie Getränke zu. Diese Roboter haben alle etwas gemeinsam: Sie sind hochspezialisiert auf ihre Aufgabe. Ihre Fähigkeiten wurden aufwändig manuell programmiert oder die Roboter wurden sogar vollständig für ihre Aufgabe entwickelt. Sollen Roboter jedoch universelle Helfer werden, müssen sie in der Lage sein, vielseitige Aufgaben zu erledigen, sich in neuen Umgebungen zurechtzufinden und idealerweise Lösungen für die Bearbeitung neuer Aufgaben zu lernen. Eine Möglichkeit, neue Lösungswege für Aufgaben zu lernen, ist den Menschen bei der Bearbeitung der Aufgaben zu beobachten und diese auf den Roboter zu übertragen. Allerdings erfordert Lernen aus Beobachtung des Menschen ein Verständnis der vorgeführten Demonstrationen und eine adaptive Repräsentation der demonstrierten Aktionen. Der Raum der demonstrierten Aufgaben ist aufgrund seiner kontinuierlichen Natur und aufgrund seiner vielen kombinatorisch möglichen Relationen zwischen Objekten und Aktionen zu groß, um in diesem Raum effektiv lernen zu können. Komplexe Aufgaben, wie beispielsweise das Zubereiten einer Mahlzeit, können jedoch als eine Aktionsfolge repräsentiert werden, die aus einer festen, grundlegenden Aktionsmenge stammen. Diese Aktionsfolge gilt es, aus der Beobachtung zu gewinnen und in einer neuen Situation zu reproduzieren. Um eine gelernte

Aufgabe in einer neuen Situation erfolgreich ausführen zu können, ist eine unveränderte Imitation der beobachteten Aktionsfolge nicht immer zielführend. Die verwendeten Objekte können sich zum Beispiel an veränderten Orten befinden oder nicht mehr vorhanden sein. Daher muss die Parametrisierung der Aktionsfolgen jederzeit an die aktuelle Situation angepasst oder es müssen neue Aktionsfolgen generiert werden können, um das Ziel der Aufgabe zu erreichen.

Die vorliegende Arbeit stellt einen vollständigen Ansatz zum Lernen von Aktionsfolgen aus Demonstrationen von Manipulationsaufgaben, sowie deren Ausführung durch einen humanoiden Roboter vor. Der Ansatz lässt sich in drei Kategorien aufteilen: 1) das Verstehen von Demonstrationen durch deren Segmentierung, d. h. die Unterteilung in Aktionen, und das Erkennen dieser Aktionen, 2) die Definition einer Repräsentationsform, mit deren Hilfe Aktionen über alle Abstraktionsebenen einer Robotersystemhierarchie beschrieben werden können und 3) die zielgerichtete Ausführung von Aktionen und Aktionsfolgen durch einen humanoiden Roboter unter Berücksichtigung des aktuellen Weltzustandes.

Der aktuelle Stand der Forschung zur Segmentierung menschlicher Demonstrationen setzt zumeist auf die Analyse von aufgenommenen Bewegungsdaten oder vereinzelt auf semantische Informationen aus Bilddaten, wie z. B. Relationen zwischen Objekten. Beides reicht jedoch nicht aus, um Aktionen zuverlässig segmentieren zu können, da Bewegungsdaten häufig mehrdeutig sind und/oder nicht alle semantischen Informationen extrahiert werden können. Daher setzt diese Arbeit darauf, semantische Informationen und Bewegungsdaten in Kombination zu analysieren. In diesem Kontext wird eine neue Methode zur hierarchischen Segmentierung menschlicher Demonstrationen vorgeschlagen. Zur Gewinnung der semantischen Informationen aus den Demonstrationen werden nicht nur Bewegungen des Menschen, sondern zusätzlich auch Bewegungen der manipulierten Objekte beobachtet, um räumliche Relationen zwischen den Händen und Objekten bzw. zwischen zwei Objekten zu extrahieren. Unter der Annahme, dass Änderungen dieser

Relationen nur durch Aktionen des Menschen hervorgerufen werden können, werden Demonstrationen bei Relationsänderungen in semantische Segmente mit deren Effekt auf die Umwelt zerlegt. Diese semantischen Segmente werden in einem zweiten Schritt basierend auf ihren Bewegungscharakteristiken analysiert, um unterschiedliche Aktionen innerhalb eines semantischen Segments zu identifizieren. Das betrifft vor allem die Aktionen, deren Effekte sensorisch nicht beobachtbar sind, wie z. B. das Schütteln einer Flasche.

Zur Evaluierung wurden die mit der hierarchischen Segmentierung erzielten Ergebnisse sowohl mit manuell segmentierten Referenzdaten als auch mit Methoden des Stands der Forschung verglichen. Der Vergleich zeigt, dass die vorgestellte Methodik dem Stand der Forschung überlegen ist, da ihre Ergebnisse im Durchschnitt näher an den manuell erstellten Referenzsegmentierungen liegen. Die extrahierten Segmente sind mit symbolischen Aktionen wie z. B. Greifen vergleichbar und können anhand des veränderten Weltzustandes und der Bewegungscharakteristik diesen Aktionen zugeordnet werden. Die extrahierten Aktionen werden zu einer Aktionsfolge zusammengesetzt und im Gedächtnis des Roboters als neue Aktion für den Planer des Roboters gespeichert.

Die Ausführung komplexer Aufgaben durch einen Roboter benötigt eine Repräsentationsform, die auf symbolischer Ebene die Verkettung von Aktionen zu Aktionsfolgen erlaubt und die gleichzeitig die Detailinformationen zur Ausführung durch den Roboter beinhaltet. Um diesen Anforderungen gerecht zu werden, wird in dieser Arbeit eine Erweiterung des Statechart-Ansatzes vorgestellt, die es ermöglicht, Aktionen auf symbolischer und sensomotorischer Ebene hierarchisch zu repräsentieren. Hierzu wird der Statechart-Formalismus um eine transitionsbasierte Datenflusskontrolle erweitert, die es ermöglicht, Fähigkeiten an die aktuelle Situation zu adaptieren und den Einsatz dieser für mehrere Robotertypen zu generalisieren.

Die Ausführung komplexer Aufgaben in sich dynamisch verändernden Umgebungen erfordert eine kontinuierliche und konsistente Wahrnehmung der Umwelt, um Aktionen und Aktionsfolgen auf die aktuelle Situation anpassen zu können. Stellt man die Welt als kontinuierlichen Raum aus Objektpositionen und Roboterposen dar, ist die Anzahl an möglichen Konfigurationen aufgrund der hohen Dimensionalität zu groß, um darin effizient nach Lösungen für Aufgaben zu suchen. Daher bietet es sich an, den Raum in eine symbolische Repräsentation zu transformieren und diesen nach Lösungen zu durchsuchen.

Im Rahmen dieser Arbeit wird das sensorische und subsymbolische Wissen des Roboters über sich und die Umwelt in eine symbolische Darstellung transformiert, um vorgegebene Aufgaben mit beobachteten Aktionsfolgen oder mit einem Planungssystem zu lösen. Eine gefundene Lösung wird unter Verwendung der vorgestellten Aktionsrepräsentation ausgeführt, kontinuierlich mittels der Weltzustandswahrnehmung überprüft und gegebenenfalls korrigiert.

Acknowledgement

First of all, I would like to thank my doctoral adviser Prof. Tamim Asfour for the opportunity to enter the field of humanoid robotics research and to pursue my Ph.D. at the laboratory High Performance Humanoid Technologies (H^{2}T). His passion for and commitment to humanoid robotics sparked also my interest in robotics and his advice and support in various ways have been invaluable. His laboratory High Performance Humanoid Technologies (H^{2}T) with its ARMAR robot family and great colleagues gave me the rare chance to work on interesting and important challenges for today's society. I would also like to thank Prof. Florentin Wörgötter for joining the committee as co-supervisor and his kind feedback.
Furthermore, I thank all colleagues at the H^{2}T for the enjoyable work environment and the great results we achieved as a team. In particular, I would like to thank Nikolaus Vahrenkamp who always helped and supported me with his long experience and calm view on everything. I would like to thank Manfred Kröhnert for being a pleasant office-mate and a good friend as we worked countless hours on ArmarX or flying RC planes during our rare spare time, which was a welcome change to working on the thesis. I thank David Schiebener for the fruitful teamwork on the long weekends working with ARMAR and the enjoyable visits to Italy to also teach the iCub some moves. With Peter Kaiser I shared memorable visits to the Humanoids conference and exploring the conference cities. I would like to thank him for his untiring help with the various projects demonstrations although he was not even involved in the project. My gratitude goes to Martin Do for his fresh ideas regarding learning from demonstration, having always an open ear for my

problems and his cheerful attitude. I would like thank Simon Ottenhaus for the teamwork on the ArmarX statechart framework and many other parts of our robot software framework as well as Ekaterina Ovchinnikova for her dedication and collaboration in the Xperience project.
Finally, I want to thank my parents for supporting me in every way my whole life and my girlfriend Mara for standing by my side and for always supporting me, especially in the stressful phases.

Contents

1 Introduction

Robotics has the potential to become one of the key technological advancements of the 21st century and to substantially improve the quality of life by transferring repetitive, tedious and hard labor tasks to service robots. In industrial environments, such tasks are already dispatched to robots due to their endurance, precision and high repeatability. Yet industrial robots are not suited to be utilized in human environments. For robots to be useful in our household everyday environment, they have to be able to navigate, to manipulate, to interact and, which is the greatest challenge, to adapt to new situations. Those environments are tailored to the requirements and capabilities of humans. Humanoid robots are well suited to be used in such environments because they are designed to be similar to humans in shape and functionality. This decreases the required effort to transfer household tasks to robots and avoids the adaption of the environment to the robot. However, equipping a humanoid robot with abilities to act in a household environment is still a difficult challenge since the robot must not only execute pre-scripted skills as used in industrial applications, but needs to adapt to the current state of the environment.

1.1 Motivation and Problem Statement

Humans have the ability to learn from their experience and to apply the acquired skills in a potentially infinite number of diverse and previously unseen situations. For example, humans can easily orient themselves in a new unseen kitchen and locate most of the ingredients and tools required for preparing a meal. Furthermore, humans are able to improvise and replace missing objects with alternatives by evaluating their properties such as shape, taste or functionality. These abilities require an understanding of the environment, advanced reasoning capabilities as well as precise manipulation skills.

One approach to equip robots with manipulation skills, which has predominantly been employed for the last decades, is to program robots manually. However, this approach requires expert knowledge and is not easily transferable to new tasks. Another approach that has gained a lot of attention in the last decade is called Programming by Demonstration (PbD) or Learning from Demonstration (LfD). The key idea here is to exploit the fact that humans already know how a task is to be performed and to transfer this knowledge to the robot. This is similar to the learning process of children, who gather parts of their knowledge by observing their parents. A main benefit of this approach is that untrained users can teach new abilities to robots instead of relying on robotics experts to program a new ability tailored to the task.

However, this "programming" approach is from the point of view of robotics far more difficult to implement. The LfD approach poses the following challenges: First, the robot has to be able to observe a demonstration and to extract the relevant parts with respect to motion, sensory information and objects. Second, it needs to transform the perceived information to its own embodiment, called *correspondence problem*, and it has to store the information in a generalized representation learned from multiple demonstrations. Third, the robot needs to apply the observed demonstration in new situations. All these challenges are not generically solved to this day and, thus, the entire learning approach is not solved either.

LfD can be divided into two categories: 1) *action learning*, i.e. learning every aspect of a single action such as *grasping an object* or more generically speaking learning of control policies and 2) *task learning*, i.e. learning how to apply known actions to solve a new complex task or in other words learning of semantic concepts. This thesis focuses on task learning.

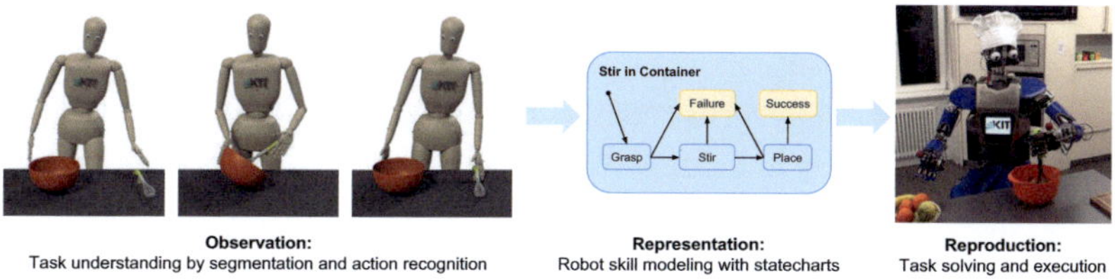

Figure 1.1: Overview of the contributions of this thesis.

Although, task learning does not imply to learn control policies for the execution of actions, still many problems remain to be solved. These problems can again be divided into three categories: understanding of a demonstration, representation of the actions used to solve tasks and the successful execution of tasks.

An observation of a demonstration is a seamless stream of motions without any known meaning. One approach to understand a demonstration is to extract meaningful segments and to associate them with already known actions. For example, preparing dough consists of *grasping*, *placing*, *pouring* and *stirring*.

The representation of the actions needs to be adaptable and hierarchical so that it can support parametrization and composition of simpler actions to more complex actions. For example, opening a door consists of grasping a handle, pulling the handle and releasing the handle.

Furthermore, the robot needs to be able to reason about the current situation and to adapt the observed demonstration in order to achieve a successful execution, e.g. the milk bottle might have been on a table during demonstration, but it might by in the fridge when the robot has to execute the task. Since successful execution cannot be ensured in an uncertain environment, failure of execution such as dropping an object during grasping needs to be detected and corrected.

1.2 Contributions

This thesis presents a novel approach for learning and for execution of object manipulation tasks by humanoid robots. This approach extends the state of the art in several points. The contributions can be divided, like most LfD approaches and as shown in Figure 1.1, into three categories: understanding of task demonstrations, representation of robot skills and execution of tasks in dynamic environments. The contributions in these categories are:

Task Understanding by Segmentation and Action Recognition
A novel hierarchical task segmentation method and a novel action recognizer are presented to segment demonstrated tasks into atomic actions. Contrary to state of the art approaches, semantic information based on object contact relations as well as motion characteristics are used for segmentation and action recognition. In both cases, the performance is improved by leveraging the orthogonality of the feature spaces.

Robot Skill Modeling with Statecharts
An extension of an existing statechart concept addressing requirements of robotics for modeling of heterogeneous robot skills on all abstraction levels is proposed. While approaches known in the literature focus on coordination, the new statechart concept also considers the data flow, which is essential for the adaptation of skills and the composition to more complex skills.

Task Solving and Execution
The third contribution allows to solve and to execute complex tasks in dynamic environments. The currently perceived state of the environment is converted online into a symbolic representation for symbolic planning. In contrast to related approaches, the proposed approach provides suggestions for object locations and object replacements based on various multi-modal strategies.

1.3 Overview

This section provides a more detailed overview of the proposed learning from demonstration approach and explains how the previously described challenges are addressed. As described before, the presented approach is divided into three parts: Observation and understanding of demonstrations, representation of robot skills and the execution of tasks. Figure 1.2 provides an overview of the structure and the contributions of the thesis.

Task Understanding by Segmentation and Action Recognition

In this thesis, demonstrations of tasks are observed with a marker-based motion capture system and converted into a normalized reference model of the human body called Master Motor Map (Terlemez et al., 2014). The demonstrations contain full body motions of a human and 6D motions of all objects. The recordings of the demonstrated tasks are taken from the publicly available KIT Whole-Body Human Motion Database presented in (Mandery et al., 2016b) containing a large variety of activities and tasks. These captured demonstrations are divided into meaningful segments by a hierarchical segmentation algorithm, which consists of two levels: The semantic segmentation on the top-level using changes in object contact relations and the motion characteristic segmentation on the bottom-level, which inserts new segments when the motion characteristic changes significantly, e.g. in case of a switch between a periodic and a discrete motion. The extracted segments serve as input to an action recognizer. In line with the segmentation, the recognition uses semantic as well as motion features. The semantic feature descriptor is an encoded version of the semantic world state used for the segmentation. It consists of object contact relations from the point in time where the segment starts and the effects the action has on the world state represented by a delta world state. The motion based part of the action descriptor uses global motion features that describe the motion characteristics of the whole action such as the periodicity of the motion. Both feature descriptors in combination provide a descriptive and distinct action descriptor. A labeled action dataset is used to train a decision tree classifier, which is able to learn the relevant features for each action. The approach is evaluated on a dataset containing various everyday activities such as preparing dough or wiping a table. A segmented and labeled task can then be saved as a new planning operator in the robot's memory.

Robot Skill Modeling with Statecharts

Since tasks are segmented into action sequences, a representation for these actions, also known as skills, is required to execute them on a robot. The requirements of the skills are manifold with respect to the used sensor and actor modalities as well as the employed algorithms. For example, the *move* skill requires self-localization, navigation algorithms and control of a holonomic platform whereas the *grasp* skill relies on object localization, grasp knowledge, visual servoing and more. However, the amount of required actions is finite. According to the study conducted by (Wörgötter et al., 2013) only 27 basic actions are needed to compose most manipulation tasks. Figure 1.3 shows ARMAR-III (Asfour et al., 2006) while executing *grasping a cup*, *pushing a chair*, *placing a cup* and *closing a fridge*. Such heterogeneous skills are difficult to learn from demonstration from the ground up. In this thesis, a powerful and flexible modeling approach is presented, which extends the finite state machine formalism called statecharts from (Harel, 1987) to the requirements of robotics. Statecharts allow hierarchical system modeling, which is crucial for robotics since primitive abilities are often combined to more complex abilities. The statechart approach is extended with several new features, most importantly transition based data flow, which is needed to design flexible and reusable robot skills. A graphical statechart editor tool is provided that enables a robot developer to design the control and data flow of new skills in a graphical way. The usability and versatility of the approach is shown in several use cases such as grasping and bipedal walking.

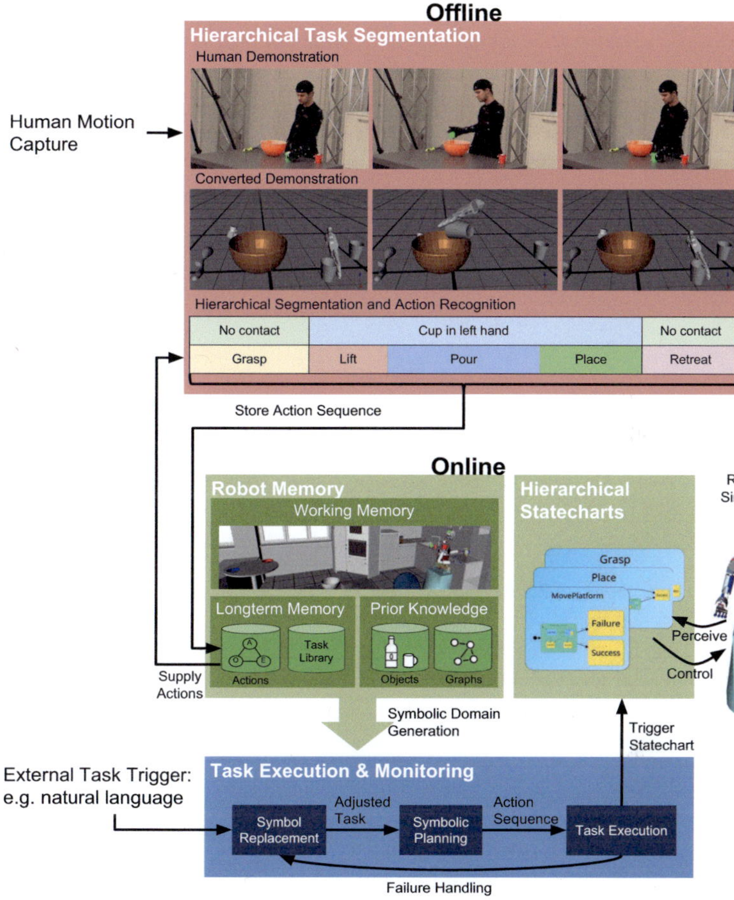

Figure 1.2: Overview of the proposed task learning approach. The red part shows the segmentation and recognition of a human demonstration, the green part illustrates the knowledge representation of the robot and the blue parts contain the components used for task execution.

Figure 1.3: Humanoid robot ARMAR-III executing different tasks in a kitchen environment.

Task Solving and Execution

To solve a task in a new environment, the robot has to be able to adapt the learned task knowledge to this particular environment. For the execution of a task in a different environment the perception results of the robot are converted into a symbolic representation consisting of predicates to be used by a symbolic reasoner. In this thesis, a classical symbolic planner is employed. Raw and processed sensor data, e.g. a 6D object localization result, is mapped to symbolic predicates stored in the memory of the robot. Based on these predicates and based on a set of actions with preconditions and effects, a symbolic domain is dynamically generated. To handle situations with objects about which the robot has incomplete knowledge, an approach called symbol replacement is used to generate replacement hypotheses. For example, the robot is asked to bring some orange juice to the human, but it does not know how to grasp the orange juice. Based on known affor-

dances of orange juice, the robot can suggest an alternative, e.g. lemonade, of which all required information is available to the robot. This approach uses multi-modal strategies based on data-mining, visual shape assessment and experiences from previous executions to suggest suitable replacements for the current task. Since classical planners always need complete domain knowledge, the dynamically generated planning domain needs to satisfy this requirement. However, objects that have not been previously seen do not appear in the robot's representation of the current environment. Linear plans, as generated by the employed planner, cannot contain sensing actions, e.g. to find these objects, because the sequence of actions depends on the result of the sensing action. Therefore, the proposed symbol replacement is also used to suggest or replace the location of objects as a hypothesis. The execution of the generated plan by a robot can fail because of a wrong world state estimation, because of failure during the execution or because the world state was changed by the human after the robot's planning. Therefore, the execution is continuously monitored and corrected through replanning if necessary.
The approach is evaluated in a kitchen scenario, in which several tasks such as *setting a table* or *preparing a salad* have to be solved. The same scenario is also used in a user study that explores whether untrained users are able to instruct the robot to solve the tasks.

1.4 Outline

The remainder of this thesis is divided into seven chapters. Chapter 2 introduces background concepts used in this thesis. Chapter 3 presents the current state of the art for segmentation of observed demonstrations, robot skill modeling, task execution and end-to-end task learning from demonstration approaches. In the following chapters, the main contributions of this dissertation are described. In chapter 4, the developed hierarchical task segmentation and action recognition approaches are presented. The robot skill modeling approach using statecharts is introduced in chapter 5 followed by the presentation of the task solving and execution in chapter 6. Chapter 7 evaluates the proposed approaches and chapter 8 discusses and concludes the thesis.

2 Fundamentals

In this chapter, essential and underlying concepts are explained, which serve as the basis for understanding the developed methods in this thesis. First, the general concept of task segmentation is introduced in section 2.1. In section 2.2, the concept of Object-Action Complexes is explained, which is employed for the execution of tasks presented in chapter 6. The original statechart concept is introduced in section 2.3, followed by an introduction into the robot development environment ArmarX in section 2.4. Finally, the fundamentals of classical symbolic planning are briefly described in section 2.5.

2.1 Task Segmentation Fundamentals

Motion segmentation refers to the process of dividing a motion sequence into several segments. The meaning of each segment depends on the application and differs between the approaches. Some approaches are interested in activities like walking and dancing (Barbic et al., 2004) while others are interested in repetitions of a rehabilitation exercise (Lin et al., 2014).

Demonstrations are usually represented as discrete sequences of motion frames, where each frame describes the momentary poses of all tracked agents and objects. Segmentation algorithms try to find the frame at which one segment ends and the next segment begins. These frames are called *key frames*. Depending on the application and algorithm, the semantic meaning of a key frame differs. In the first example of this section, key frames are defined when the motion style changes while in the other example key frames are defined when a new repetition starts, which heavily depends on the exercise (see Figure 2.1 for an illustration).

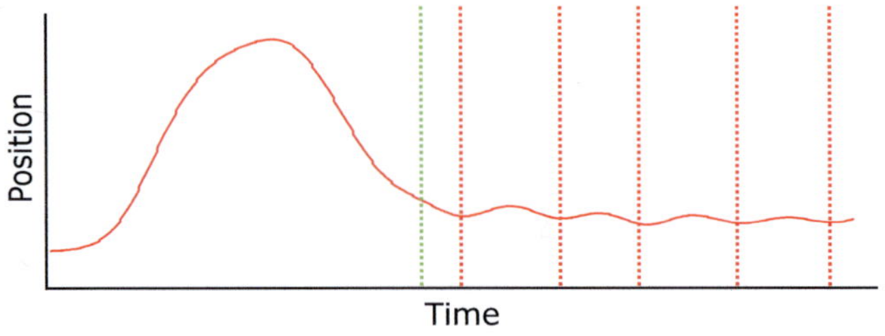

Figure 2.1: Differences between segmentation algorithms: The green line could represent a key frame detected by an algorithm extracting different types of motions whereas the red lines could represent key frames detected by an algorithm extracting repetitions of an exercise.

Task segmentation is an alteration of motion segmentation and refers in this thesis to the process of dividing a demonstration into single actions which change the semantic state of the environment. Since semantic state changes are not reflected in the motion of the agent only, additional information should also be recorded during the demonstration. Different approaches use additional sensors like haptic or force sensors (Jäkel et al., 2011) or track also the objects in the scene (Aksoy et al., 2010; Ramirez-Amaro et al., 2014) as it is done in the approach in this thesis. This additional information is used to extract semantic relations between objects (Aksoy et al., 2010; Ramirez-Amaro et al., 2014) or constraints for the motion (Jäkel et al., 2011). Task segmentation aims at extracting actions that can be directly mapped onto a robot (e.g. (Aein et al., 2013)). Thus, the result of a task segmentation is a sequence of primitive actions that in this particular succession represents the demonstrated task.

2.2 Object-Action Complexes

Object-Action Complexes (OACs) were introduced by the PACO-PLUS project[1] and presented in (Wörgötter et al., 2009), (Krüger et al., 2011) and (Geib et al., 2006) as a formalism for sensory-motor processes to capture the interaction between objects and actions. OACs are designed to achieve adaptive and predictive behavior on all abstraction levels. Each OAC is defined over an attribute space that can be perceived by the robot itself and is changed by the actions. In section 6.4 the process of translating noisy, continuous and uncertain sensor data to a symbolic domain is described.

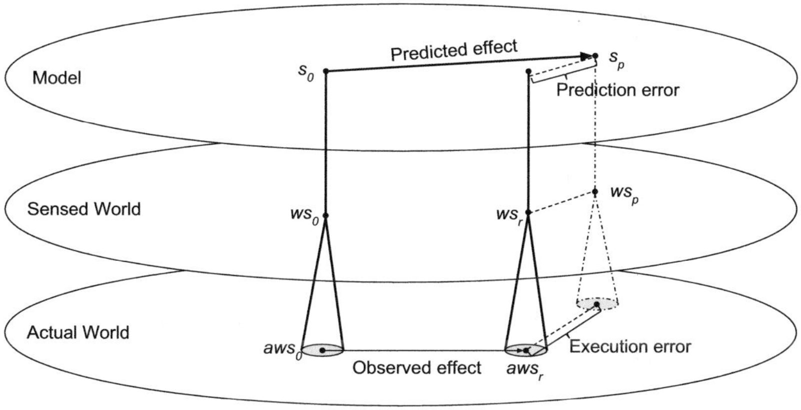

Figure 2.2: OACs work in three different attribute spaces: Actual world space, sensor world space and model space. A state from the actual world (*aws*) is perceived with noisy sensors and mapped into the sensed world space (*ws*). The sensed world state can be mapped without loss into the model world state (*s*). The execution of an OAC might lead to a different world state than the predicted world state. Hence, the OAC should learn to minimize the prediction error as well as the execution error.

An OAC consists of the triple: a link to an execution control program, a prediction function and a success measure. The execution control program is responsible for executing the action on the robot. This is done in this

[1] `www.paco-plus.org`

thesis with the proposed statechart approach (chapter 5), which provides the requirements for modular and hierarchical skills for any abstraction level.
The prediction function is responsible for predicting the effects the associated execution of the control program will have on the defined attribute space. In this thesis, this function predicts the change of the symbolic state of the environment that the action will cause if executed successfully.
The success measure stores the success rate of an OAC over time. In this thesis, the success of each OAC is measured by evaluating the final event of each statechart after execution, which signals whether the statechart itself deemed the execution as successful, and by comparing the expected outcome (prediction function) of an OAC with the perceived outcome.
To be able to learn from execution trials, each run is stored as an *experiment* with the triple of the state before execution, the predicted state and the perceived state after execution. OACs use these experiments to improve their prediction function to reduce the discrepancy between the predicted outcome of an OAC execution and the observed outcome. Furthermore, the experiments are also used to adapt the control parameters to reduce the execution error, which is the error between the actual execution result and the predicted execution result. Depending on the OAC it can either be easier or make sense to reduce the prediction error or the execution error. Figure 2.2 illustrates the different attribute spaces and mappings. Since these learning tasks are highly dependent on the specific OAC, the decision which learning algorithm is to be used is up to the OAC designer.

2.3 Original Statecharts

(Harel, 1987) proposes a new formalism to represent and describe complex systems. The limitations of finite state machines (FSM, (Gill, 1962)) motivated him to develop the more powerful *statecharts* representation. Like FSMs, statecharts consist of states and transitions between these states, but Harel introduces several new notation features compared to FSMs. The main aspect of statecharts is to reduce the complexity for the human developer. First of all, Harel introduced a *hierarchy* of states, meaning states can contain a full statechart themselves allowing for reuse and composition of states. However, these state levels are not completely separated from upper or lower levels, since he proposed *inter-level-transitions* to directly jump into substates. Furthermore, *orthogonality* is used to parallelize the execution of different states on one statechart level. Each orthogonality level of one state is executed in parallel, independently of the other levels. A *history*-connector is added to give states a memory, controlling which substate is entered when a state is reentered. Based on the fulfillment of conditions, *Condition*-connectors control which state a transition leads to. Finally, each state can be connected to actions, being triggered during different phases of the state: entering, leaving and an action that is executed repeatedly as long as the state is active.

2.4 The Robot Software Development Environment ArmarX

ArmarX[2] (Vahrenkamp et al., 2015) is the robot software development environment used and significantly extended in this thesis. It is the base for every component related to robot execution (realization of chapter 5 6). It is used for connecting to the robot memory and for convenient communication between different programming languages in the implementation of chapter 4.

In contrast to YARP (Metta et al., 2006) and ROS (Quigley et al., 2009) ArmarX provides not only middleware functionality for distributed processing, but a complete cognitive robot architecture designed for complex robot systems such as humanoid robots. The middleware functionality of ArmarX is an extension of the mature middleware ZeroC Ice (Henning, 2004) in order

[2] ArmarX: `https://armarx.humanoids.kit.edu`

to handle dependencies between components and their life cycle. On top of the middleware, the framework consists of three layers (see Figure 2.3): the low-level drivers and control, the mid-level robot capabilities and the high-level robot coordination.

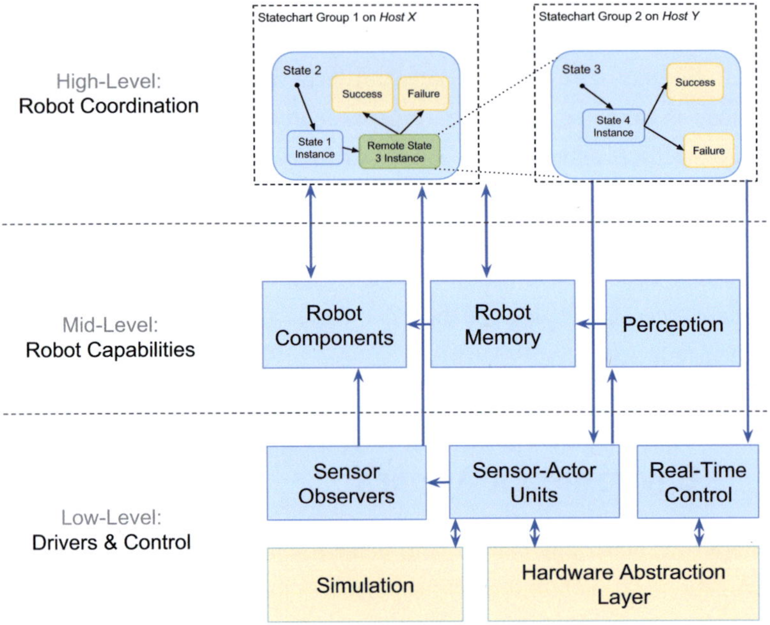

Figure 2.3: Layers and main elements of ArmarX.

The *low-level* layer unifies the access to the hardware or simulation as well as offers all sensor data in a generic format to all interested components as sensor observers. These sensor observers are used by the statechart approach to trigger conditional events (see subsection 5.3.5) based on condition checks located directly at the source to minimize delay and data traffic. Additionally, real-time controllers run in this layer, i.e. as closest as possible to the hardware. These controllers can be activated and parametrized from the higher layers.

The *mid-level* layer provides services and storage for any kind of knowledge. These services range from self-localization of the mobile platform over object recognition to inverse kinematics. They offer the functionality to all components in the network. The storage is used for the robot's memory, which is essential for the realization of chapter 6. This subframework is called MemoryX (Welke et al., 2013a; Kozlov, 2013) and will be described in more detail in the subsection 2.4.1.
On start up all components are idling. The *high-level* layer brings the robot to life by coordinating the mid- and low-level components. This is realized with statecharts (chapter 5) and symbolic planning (chapter 6). Robot skills and behaviors are modeled with statecharts and combine the services to complex skills. For example, the *grasp* skill uses the visual object and hand localization, the robot memory, the autonomous gaze selection, inverse kinematics, visual-servoing and joint control.
All three layers are developed based on several design principles, which are also reflected in the statechart and task execution approach:

- *Disclosure of the internal state*: The state of the system at any abstraction level can be inspected at runtime. The meta state of system shows the state of all components and their dependencies. All sensor data is offered in a unified way. The robot's memory content is disclosed as well as visualized in a 3D view of the environment, objects and agents. Statecharts and their control and data flow as well as the state of a task can be visualized and inspected.

- *Distributed processing*: All components and statecharts are reachable via network and can effortless be distributed as needed over multiple hosts.

- *Shared and distributed resources*: The robot's memory can easily be used, modified and extended by any distributed component. Statecharts are reused as substates of other statecharts.

- *Type safe and unified interfaces*: All calls, also network calls, are type safe. Even the statechart data flow is type safe due to automatic checks in the graphical statechart editor (section 5.6).

2.4.1 *MemoryX:* Robot Knowledge Representation

In order to equip a robot with a symbolic planning system that reasons on the current knowledge of the robot about the environment, a powerful robot memory system is needed. This memory system needs to store information about prior knowledge, which the robot cannot learn by itself and which was provided by the developer. Further, the current world state needs to be appropriately represented and continuously updated, containing among other entities object knowledge, agents, environment maps, affordances and relations between entities of the memory. Last but not least, the robot needs to have the possibility to update and persistently store data that it acquired by itself and that can be useful in the future, such as success rates of OACs or the currently perceived body schema (Ulbrich et al., 2012).
To this end, the memory system MemoryX of ArmarX (Welke et al., 2013b; Kozlov, 2013) is used for the proposed task execution approach and extended to fulfill these requirements.
MemoryX consists of three memory types: Inspired by the findings of Atkinson and Shiffrin (Atkinson and Shiffrin, 1968) from the neuroscience field, it contains the *Working Memory* and the *Long-term Memory*. The third memory type, *Prior Knowledge*, is designed to store knowledge provided by the human. While *Long-term Memory* and *Prior Knowledge* are persistent, the *Working Memory* is volatile.

Internal Structure of the Robot Memory

Each memory is organized as a set of segments. Each segment can only contain one specialized type of data, e.g. object types or OACs, but an arbitrary number of instances of these. These specialized types are type safe convenience views on the basic generic element `Entity`, which is the core construct of the memory. An `Entity` consists of a name, of a unique id, of relations to other entities and of a list of `Attributes`. An `Attribute` consists of a list of variant data types that can morph into any data type and of a probability distribution for this attribute. The relations to other entities allow specifying hierarchies in memory segments. Figure 2.4 shows the structure of one memory in MemoryX.

Prior Knowledge

The prior knowledge memory data is handcrafted and contains two memory segments: An object type segment and an environment navigation graph segment. Each object type contains several attributes:

- 3D mesh that represents the object type visually
- object recognition descriptor, e.g. for textured or shape based object recognition
- grasp definitions that provide information about how to grasp an object
- motion model that predicts how an object moves between localizations
- place orientation that provides a hint how an object should be placed on a surface

The navigation graph segment consists of graph nodes with the following attributes:

- list of nodes that can be reached from this node
- 6D pose
- node type for the symbolic planner to give nodes different semantic meanings, e.g. a location on a table where objects might be or a landmark for robot navigation

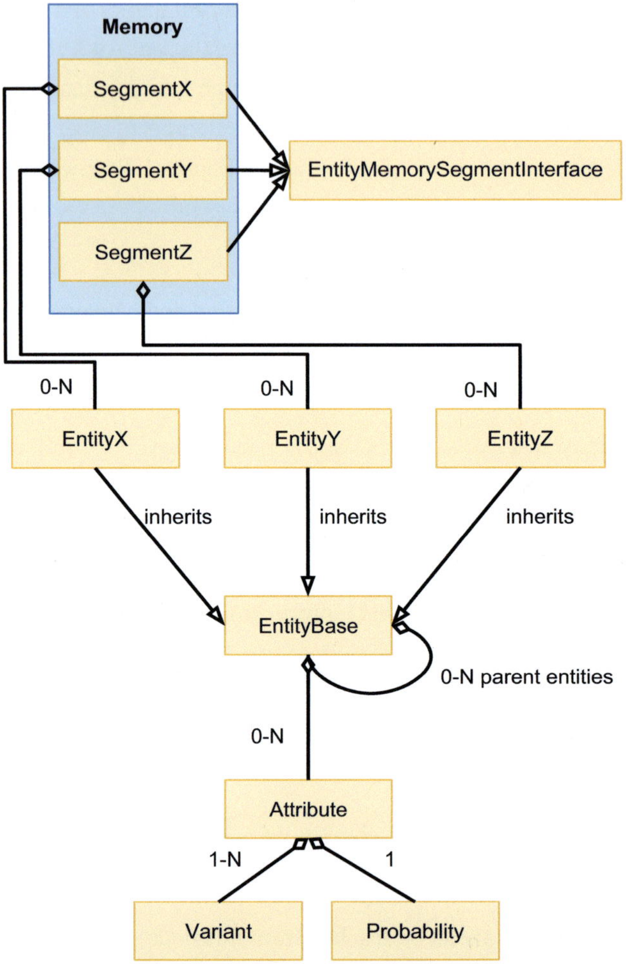

Figure 2.4: A memory in MemoryX is a segmented memory. Each segment contains data elements of one type (e.g. object class). Each element type is of the base type `Entity`. An entity is a generic data type with a set of attributes. Attributes themselves consist of $1 \cdots n$ values of a generic type called `Variant` and a probability measure. Variants can be any type of data, from primitive types like `int` or `string` to arbitrary complex types like 6D-poses or n-dimensional trajectories.

Working Memory

The working memory represents the knowledge of the robot about the current state of the world. It consists of several segments containing object instances, agent instances, entity relations, detected shape primitives and available affordances. The object instance segment entries consist of the following attributes:

- object type
- all attributes of the object type that it is created from
- 6D Pose
- existence certainty
- current motion model
- localization priority

The agent instance segment entries consist of the following attributes:

- 6D Pose
- robot model
- full robot state containing joint angles, joint velocities, joint torques

The relation segment entries are n-ary predicates, i.e. predicates describing a relation between n entities that can be true or false. Other segments containing information such as perceived affordances or perceived geometric primitives are not used in this thesis.

The object instance segment is enhanced with sophisticated update mechanisms. The reader may consider the case in which an object is localized by the robot at one point in time. Afterwards, the robot is looking in another direction. Will the object still be there when the robot looks in the first direction again? A human or another robot might have moved the object in the meantime. What happens to the object if another robot had the object in its hand when it was localized the last time? These are important considerations for a consistent robot memory if the robot is not able to track all objects at once. To this end, every object instance is enriched with a location uncertainty and a motion model that predicts how the object might move. Each motion model provides information how the pose and location uncertainty changes over time. Most object instances are enriched with the static motion model by default. This motion model keeps the object instance at one pose, but it increases the location uncertainty constantly over time. New localization updates are fused with a Kalman filter (Kalman, 1960) into the existing pose and decrease the location uncertainty again. Additionally, the Kalman filter also deals with noisy localization results and treats the confidence of the object localizer accordingly since state of the art object localizers frequently produce false positives or noisy poses. One false positive has only little impact if the object has already been localized several times at a similar position.

Other important motion models are the robot hand motion model and the attached-to-object motion model. The robot hand motion model moves the visually localized hand according to the kinematics of the robot. A visually localized hand is needed to deal with the error of the 3D localization, yet relative changes can be used from the kinematics. The attached-to-object motion model moves the object relative to another object that the robot tracks. This is usually used to attach a grasped object to the hand of the robot. This is especially needed because most object localizers cannot localize an object if it is occluded by a hand.

Additionally, the object instance is enriched with an existence uncertainty, which is important for the autonomous gaze selection and planning. If an object is not seen anymore at the old location, it is not immediately removed from working memory since the object localizer might have reported a false negative. The existence certainty is gradually reduced after each localization iteration if the object cannot be seen anymore at the old location.

The autonomous gaze selection keeps track of requested objects, e.g. from the task execution system, by leveraging the pose uncertainty and the existence certainty. Objects are only attempted to be localized if the robot is looking at their old position to reduce the average computation cost, unless an object has a low existence certainty, which means in other words the object could be anywhere or not there at all. Based on the pose uncertainties of all requested objects the gaze direction is chosen as presented in (Welke et al., 2013b). The algorithm directs the gaze and in turn keeps track of all objects by maximizing a score over the objects with the highest pose uncertainty, a localization priority of each object and the benefit of looking at multiple objects at once. All these mechanisms are needed in a system with partial observation possibilities to keep the working memory as consistent as possible.

Long-term Memory

The long-term memory is used to store and retrieve information that the robot acquires or refines itself through exploration and experience. In particular, for this thesis two types of information are important: OACs and common places of objects (Welke et al., 2013a; Kozlov, 2013). The OAC statistics and refinements can be stored in the long-term memory. Additionally, their parameters, preconditions and effects are stored here for the planning system. The common places are the learned knowledge about where objects have been seen in the past by the robot. This data is accumulated for each object and merged into clusters that can easily be turned into symbolic locations by the associated replacement strategy (see subsection 6.3.2).

2.5 Symbolic Planning Fundamentals

The task of generating a sequence of actions to achieve a given goal in a symbolic world is called symbolic planning. Classical symbolic planning searches for plans in worlds that are fully observable, deterministic, finite, static (change is only induced by actions of agents) and discrete (regarding time, action, objects and effects) (Russell et al., 2003). This type of planning is used in the task execution approach presented in chapter 6. There are several different languages to describe the state of the world such as Stanford Research Institute Problem Solver (STRIPS) (Fikes and Nilsson, 1971) or Action Description Language (ADL) (Pednault, 1987). The main concepts of ADL, which is used in this thesis, will be described next.

The state of such a symbolic world is represented as a conjunction of positive or negative (first-order) predicates that describe the world, e.g. $sunny \wedge open(Window)$. The state of unmentioned predicates is unknown based on the *open world assumption*, resulting effectively in a tristate of the predicates: $true, false$ or $unknown$.

A *goal* is a partially specified state, represented as conjunctions or disjunctions, e.g. $(sunny \wedge open(Window)) \vee closed(Window)$. A goal is satisfied if the goal expression validated against the current state is true.

An *Action* is defined as a quadruple: a unique name, a set of typed parameters, preconditions and effects. The unique name is the identifier of the action. The set of typed parameters is used in the effects and conditions to realize flexible actions. If actions were specified without parameters, one action would be needed for each possible configuration, e.g. $graspBottle()$ instead of $grasp(Bottle)$.

The preconditions are first-order logic expressions with only action parameters used in the predicates and must be fulfilled in order to execute an action. The effects of actions are the changes caused by the action to the state of the world. The changes can add or delete negative or positive predicates. Deleted predicates imply that the state is unknown. If the state of an predicate is known, the state is always explicitly expressed as positive or negative.

An example action could look like this:

$$\begin{aligned}
&grasp(o : Object, l : Location, h : Hand) : \\
&\qquad Preconds : empty(h) \wedge on(o,l) \\
&\qquad Effects : Add(\neg empty(h)) \wedge Del(on(o,l))
\end{aligned}$$

It is to be noted that any literal-variable combination can only appear once in the world state and adding a positive literal also invokes deletion of the negative literal and vice versa. Additionally, preconditions, effects and goals can contain quantified variables, e.g. $\exists x\ on(x, Table)$.
These are the main components of a planning domain, but additionally all types and constants present in the world must be specified. A planning domain describes and contains all information necessary about the world for the planning problem. Specifically, a planning domain consists of the domain description and the problem description. The domain description contains all actions, type declarations, predicate declarations and typed constants whereas the problem description contains the initial state of the world and the goal. A reduced, but complete domain in the PKS (Petrick and Bacchus, 2002) syntax that is used in this thesis is given in Appendix A. A domain can be used by a symbolic planner to generate a grounded action sequence that can be executed by a robot. How such a domain description is generated from the current robot knowledge and how it is utilized is described in chapter 6.

3 State of the Art

There have been numerous publications on the topic of learning from demonstration. This chapter provides an overview of the literature and the state of the art. The chapter is divided by the subproblems tackled in this thesis. First, literature dealing with task understanding by segmentation is presented. Section 3.2 describes robot skill modeling and programming approaches followed by task execution and monitoring approaches in section 3.3. In section 3.4, end-to-end learning from demonstration approaches are presented and discussed.

3.1 Task Understanding by Segmentation

One important component of learning from observation is the ability to understand the demonstration performed by the teacher. Since tasks can be understood as a sequence of actions (Dillmann, 2004), the first step to understand a task is to segment it into the actions it consists of. This is a difficult problem to solve since there is no clear ground truth of what is the correct segmentation of a task. Furthermore, the demonstration can vary between individuals due to kinematic and dynamic differences (Newell et al., 1998) and vary in each repetition since humans cannot reproduce tasks perfectly, e.g. due to fatigue (Winter, 1991). Furthermore, the correct segmentation solution depends on the target application (Lin et al., 2016b) and even then it is ambiguous at which point one segment ends and a new segment starts. For example, in (Lin et al., 2014) physiotherapy sessions are recorded with the goal to extract each repetition of each exercise, while (Lv and Nevatia, 2006) aim to extract walking or waving actions. In addition, most approaches found in the literature are tailored onto their task - especially if they are data-driven and the training data set is very specific (Lin et al., 2014). Motion segmentation has recently been of great interest in the research community although it has already started in the 90's (Kang and Ikeuchi, 1995). A large amount of different approaches has been developed, which are summarized and compared in the survey paper by (Lin et al., 2016b) or more generally about time series segmentation in (Keogh et al., 2004).

(a) (b)

Figure 3.1: Different motion capture techniques: (a) RGB based marker-less skeleton tracking. *Source:* (Azad et al., 2008) © 2008 IEEE (b) Marker-based tracking of skeleton and objects as used in this thesis. *Source:* (Wächter and Asfour, 2015) © 2015 IEEE

3.1.1 Capturing Demonstrations

Before segmenting a demonstration, it needs to be observed and recorded by technical means. The different segmentation approaches utilize various sensor systems based on their requirements and availability such as marker-based motion capture systems (VICON, 2016; OptiTrack, 2016; Motion Analysis Corporation, 2016), mobile systems based on inertial measurements units (IMU) (Roetenberg et al., 2009; Burns et al., 2010) or (depth) cameras (Han et al., 2013; Zhang, 2012; Sell and O'Connor, 2014). Figure 3.1 shows examples of different motion capture techniques.

Marker-based motion capture systems provide the highest precision of the different systems, but require attaching multiple reflective markers per body segment, e.g. 53 markers for the whole body in total in (Terlemez et al., 2014), and a set of infrared cameras around the subject. Since these systems rely on vision they are susceptible to occlusions and often require manual post-processing of the recorded data to fill gaps during occlusion of markers. Segmentation algorithms using those systems are (Ilg et al., 2004; Lv and Nevatia, 2006; Kulic et al., 2009; Kulic and Nakamura, 2010; Vögele et al., 2014; Fox et al., 2014; Mandery et al., 2016a) and the approach presented in this thesis.

IMU-based systems pose less restrictions for the usage, are not affected by visual occlusions and are easier assembled, but also less accurate and prone to position drift since they estimate the position from acceleration data with a Kalman Filter (Luinge and Veltink, 2005; Lin and Kulić, 2012). Algorithms using IMU-based systems are (Li et al., 2013; Berlin and Van Laerhoven, 2012; Chamroukhi et al., 2013; Yuwono et al., 2013; Aoki et al., 2013).

Depth cameras as sensors for tracking demonstrations are most flexible, require no preparation of the demonstrator and can be attached to a robot for online tracking. Yet they also achieve a lower accuracy and a significantly lower robustness regarding recognition errors. Additionally, complex and computationally costly algorithms are needed to extract the human pose or object pose from the captured point cloud which induces a higher latency and lower data frequency (Livingston et al., 2012). Algorithms using depth cameras are for example presented by (Aksoy et al., 2015, 2016) and (Devanne et al., 2017).

Tracking systems using (stereo) RGB cameras as described by (Azad et al., 2008) and (Bandouch and Beetz, 2009) are the least accurate, but the most natural. They are used by (Kuniyoshi et al., 1994; Bradski and Davis, 2002; Ratanamahatana and Keogh, 2004; Bashir et al., 2005; Aksoy et al., 2011; Lee et al., 2012; Pei et al., 2013; Ramirez-Amaro et al., 2015).

3.1.2 Motion based Task Segmentation

Motion based task segmentation analyzes trajectories in joint space or task space with the goal of directly finding meaningful segment points or segments. The approaches can be categorized as online or offline approaches and supervised or unsupervised approaches. Online means in this case that the computational complexity is low enough to execute the algorithm as fast as the motion data is recorded as well as that the algorithm is causal. Many approaches are acausal and need either the whole demonstration or a window of the trajectory after the current frame for their analysis. Online approaches are for example presented by (Kulic et al., 2009; Lan and Sun, 2015; Amft et al., 2005; Barbic et al., 2004) whereas the algorithm presented in (Ataya et al., 2013; Chamroukhi et al., 2013; Lv and Nevatia, 2006; Vicente et al., 2007) only support offline segmentation. Most methods in the literature perform segmentation by identifying segment points instead of recognizing segments themselves. Some of these algorithms segment by thresholding on feature vectors such as zero-velocity crossings (ZVC) (Pomplun and Mataric, 2000; Lieberman and Breazeal, 2004; Fod et al., 2002).

(Pomplun and Mataric, 2000) study how rehearsal of motions effects motion imitation and assume that motions have clear start and end poses. Other feature spaces used with zero-crossings are joint acceleration (Guerra-Filho and Aloimonos, 2007; Ricci et al., 2013), linear acceleration (Yuwono et al., 2013) and angular jerk (Yamamoto et al., 2006). Yet these algorithms suffer all from the assumption that the crossings always indicate a segment and tend to over-segment motions, e.g. a circular motion in 2D has either too many zero-velocity-crossings (on each quarter) if each dimension is considered separately or none if both dimensions are considered jointly. Furthermore, threshold based methods always need tuning, which becomes more difficult with higher dimensionality and heterogeneous features. Other algorithms segment by thresholding on distance metrics. Commonly used metrics are Euclidean distance (Mueen et al., 2009; Hao et al., 2013), Mahalanobis distance (Barbic et al., 2004) and the Kullback-Leibler divergence (Kulic et al., 2009; Kohlmorgen and Lemm, 2002).

(Vögele et al., 2014) employ a self-similarity matrix, which compares the current frame to future frames. Segments are extracted by detecting distinct temporally coherent activity segments with a neighborhood graph formed by a kd-tree (Bentley, 1975). The mentioned segments are then further analyzed to find recurring patterns, i.e. motion primitives. The idea to segment by searching for segments which are most different is similar to the idea presented in this thesis.

(Koenig and Mataric, 2006) propose using signal variance as the distance metric. Their segmentation exploits the assumption that drastic variance changes suggest that the observed action changed.

(Devanne et al., 2017) also use a statistic feature, i.e. the standard deviation of spatio-temporal shape poses. Segments are extracted by detecting local minima of the standard deviation. A low standard deviation of the shape poses corresponds to slower motions according to the authors.

Hidden Markov Models (HMM, (Rabiner, 1989)) are instrumented by many approaches since they model the probability of state sequences. This concept corresponds nicely to motion trajectories. One approach using HMMs is introduced by (Kohlmorgen and Lemm, 2002), which was later extended with clustering by (Kulic et al., 2009). State transitions of the HMM signal new segments in these approaches. (Kulic and Nakamura, 2010) and (Aoki et al., 2013) use hierarchical HMMs for online and incremental segmentation of human behavior. (Asfour et al., 2008) present an approach for learning dual-arm tasks with HMMs consisting of key frames extracted in a preprocessing step. These key frames are defined based on direction changes of single joints if sufficient time has passed since the last key frame and the joint angle difference is sufficiently high. Additionally, key frames are inserted after every pause of motion. By using only the key frames for training of the HMMs a high number of states is avoided, which is beneficial for matching of multiple demonstrations.

Another type of approach is template matching (Kang and Ikeuchi, 1995; Zhang, 2012; Ormoneit et al., 2001; Lin and Kulić, 2014; Lee et al., 2015), in which patterns are fitted over windows of the motion to determine segments. (Ormoneit et al., 2001) examine in a first step the signal-to-noise ratio to adjust the window size used for fitting curves of premade templates. (Lin and Kulić, 2014) use velocity sequences in combination with acceleration crossing points to locate segment points in addition to a fine-tuning step with HMMs to avoid over-segmentation. (Meier et al., 2011) match Dynamic Movement Primitives (DMP) from a predefined primitive library on (partially) observed motions to find segments and recognize the actions. A drawback of template matching is that it always requires to generate a library of templates, which only covers all possible motions if applied in a restricted use case such as segmentation of rehabilitation exercises. Furthermore, variability in the execution is not always well enough generalized by the templates (Lin and Kulić, 2014).

(Beaudoin et al., 2008) present a motion-motif finding algorithm based on strings constructed from a pose-alphabet, where each motion pose is labeled with a letter. In combination with an adjacency matrix the distance between letters and in succession strings is utilized to cluster similar motions and to create a motion-motif graph, which can be used for segmentation.

A popular unsupervised approach is presented by (Barbic et al., 2004) based on Principal Component Analysis (PCA) and as an improvement based on Probabilistic Principal Component Analysis (PPCA, (Tipping and Bishop, 1999)). They employ the paradigm that the complexity of motions is different and when the complexity exceeds a defined level a key frame should be inserted. In a more technical wording, the segmentation method generates key frames based on the gradient of the error of a motion frame after reconstruction from a low dimensional representation created with PCA/PPCA. The reconstruction error is created for every frame of the motion and used to calculate the gradient by comparing the current error to the error of a frame half a second earlier to compensate for noisy data. A new key frame is extracted if the difference between the current error gradient and the average gradient error of all previous frames is three times higher than the standard deviation of all previous frames. After a new key frame has been extracted, the algorithm starts again with the rest of the motion. This method is then also extended to use PPCA instead of PCA.

(Lv and Nevatia, 2006) present an approach based on HMM and multi-class AdaBoost, in which several different features consisting of combinations of related joint coordinates are used to generate a 141-dimensional feature space that covers different aspects of the motions. For each feature and for each motion class the motion dynamics are learned by one HMM. Given a motion sequence, the observation probability is calculated with each HMM. Each HMM itself does not provide a reliable classification. Thus, AdaBoost is employed to combine the multitude of weak classifiers to a more reliable classifier.

Another unsupervised approach, which gained a lot of attention recently, is proposed by (Zhou et al., 2013) based on clustering. Figure 3.2 shows the approach schematically. They present an extended version of k-means clustering (MacQueen, 1967) called Hierarchical Aligned Cluster Analysis (HACA), which clusters motion segments based on their frame-wise similarity. Kernel functions are used to overcome the problem of learning only spherical clusters. To account for variances in segment length, an extension of Dynamic Time Warping (Yi et al., 1998), called Dynamic Time Alignment Kernel, is applied that assigns a mapping for each frame of two segments and thus makes two segments with different lengths comparable. The hierarchical extension of ACA uses multiple levels with different temporal resolutions resulting in different lengths of the segments. This reduces the computational complexity and provides a hierarchical decomposition at different time scales.

One drawback of this approach is that the exact number of clusters, i.e. the number of different motions, needs to be known in advance for each motion sequence that is to be analyzed. In other words, a human has to extract the motion types contained in a motion sequence and parameterize each trial by hand. This limits the usability to specific segmentation applications, such as analyzing a specific exercise, and makes the approach not applicable on a general motion database. In combination with the problem of motion variances, such as grasping objects at different places, the usage of this approach requires tuning of the parameters for each dataset.

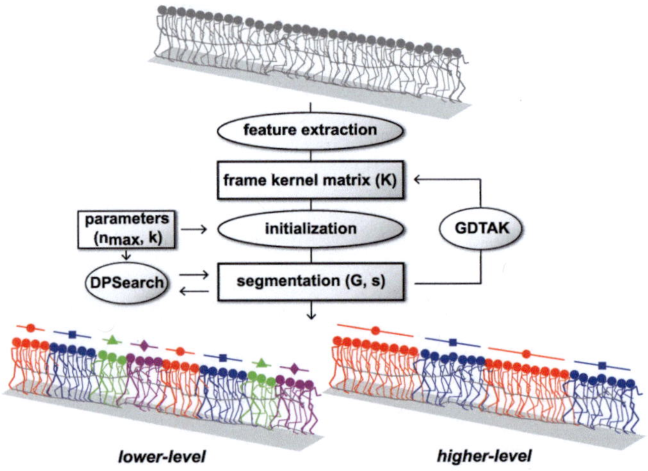

Figure 3.2: Hierarchical Aligned Cluster Analysis used to segment motions into recurring motion primitives. *Source:* (Zhou et al., 2013) © 2013 IEEE

(Lin et al., 2014) propose a supervised, data-driven approach for segmentation of physiotherapy exercises. Their approach classifies each frame of a motion sequence into either of two classes: a segment point or a non-segment point. Figure 3.3 shows an example of the input data for the classification. Points around a transition between two segments are labeled as segment points and the remainder are labeled as non-segment points. If they had only used the exact transition points as segment points, the imbalance between non-segment points and segment points would be too high for a classifier to learn the problem. Since many machine learning algorithms are interface-wise interchangeable, the authors evaluated three different learning algorithms with different combinations of pre- and post-processing. As an optional pre-processing step PCA or Fisher's Discriminate Analysis (Jain et al., 2000) is utilized. As binary classifier one of the following algorithms is used: k-Nearest Neighbor (Jain et al., 2000), Quadratic Discriminate Analysis (Jain et al., 2000), Radial Basis Functions (Orr, 1996), Support Vector Machines (Burges, 1998) or Artificial Neural Networks (Jain et al., 1996). The authors evaluated boosting (Freund et al., 1999) and bagging (Breiman, 1996) as post-processing options.

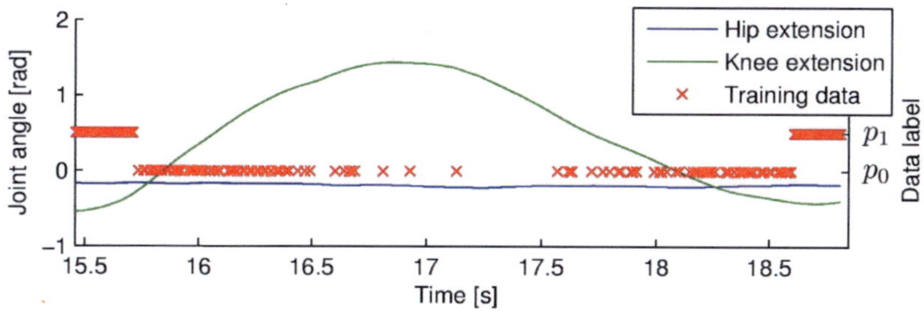

Figure 3.3: Segmentation based on data point classification. p_1 denote segment points, while p_0 denote non-segment points. Instead of only using single segment points for classification a window around each segment point is used to obtain a more balanced training set. *Source:* (Lin et al., 2014) © 2014 IEEE

To consider not only one frame for classification they employ feature stacking, i.e. positions from a few frames before and a few frames after the current are added to the feature vector. Since the classification label distribution is not balanced the classes are downsampled with points preferred that are close to segment borders. With this approach, the authors achieve impressive results with classification rates of up to 100 %. However, they only evaluate the approach on simple motions like squatting with only three IMU sensors attached to the subject, which were also cleanly executed if the example shown in the paper is representative. How well this algorithm performs on a larger dataset with higher variances in motion execution is unfortunately not evaluated. (Lin et al., 2016a) propose to utilize the fact that human motion is optimized in regard to different criteria which depend on the executed action. They extract from motion data which criteria, e.g. maximum force output or minimal energy consumption, in a motion segment has most likely been optimized. To this end, they use a sliding window in which Karush-Kuhn-Tucker optimality conditions are minimized. Examples for optimized criteria are exerted force and angular acceleration. Whenever an optimized criteria changes, a key frame is inserted.

3.1.3 Semantic Task Segmentation

While the majority of approaches relies on motion data, some approaches utilizing semantic information emerged in recent years. (Ziaeefard and Bergevin, 2015) dedicate a survey paper to semantic activity recognition based on single images or RGB videos. Activity recognition can be similar to task segmentation: When atomic actions are to be recognized from a longer sequence some kind of decomposition is also required as a preprocessing step. (Ziaeefard and Bergevin, 2015) categorize the different semantic approaches into several feature spaces: *Poselets*, which are parts of a body posture that are distinctive for a specific action, *scene and object context*, which have a semantic relation to the actions performed, and attributes that expose more information about objects or activities.

Poselet based approaches are fairly similar to motion based template matching methods, but assign a specific semantic to parts of the body and combine multiple poselets to recognize actions (Yang et al., 2010; Maji et al., 2011; Raptis and Sigal, 2013; Wang et al., 2014).

Approaches based on objects and scene context are similar to the semantic part of the hierarchical segmentation proposed in this thesis. They try to leverage the additional information provided by the context to restrict the possible actions. (Marszalek et al., 2009) use movie scripts as annotation for videos and try to extract recurring relations between scenes and actions. (Zhang et al., 2014) separate the background from the person in the foreground and learn spatio-temporal interest points for the person and several coarse features for background regions to combine them in a bag-of-features model.

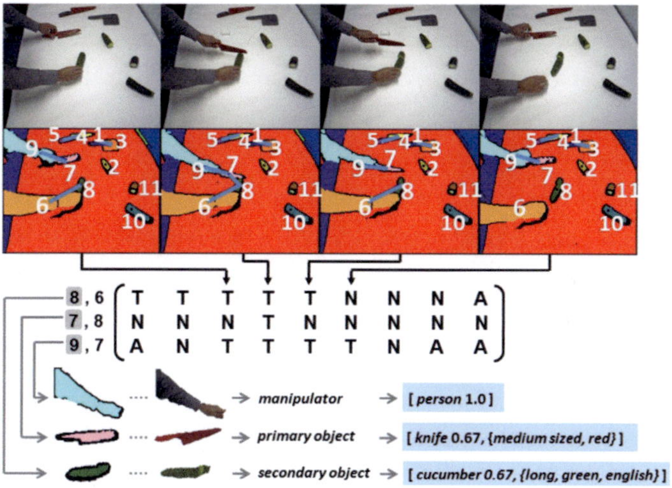

Figure 3.4: Segmentation of manipulation actions based on object relation changes extracted from color and depth regions calculated from RGB-D images. *Source:* (Aksoy et al., 2017) © 2017 IEEE

(Aksoy et al., 2010) proposed their first approach in 2010 utilizing contact relations between objects for object categorization as well as action recognition. It is a model free approach that segments an input video stream into regions based on their color and estimates contact relations based on spatial proximity of the color regions as shown in Figure 3.4. Actions are not directly recognized with traditional labels, but based on their effect on the environment. Objects are categorized by their affordance (Gibson, 1979) exhibited during the observed action sequence. They later extended this approach to RGB-D sensors (Aksoy et al., 2015). In a further extension of their work (Aksoy et al., 2016), they are able to segment and recognize actions from sequential and concurrent action sequences. Additionally, basic manipulation

primitives can be extracted as well as scene entities that share the same roles in the observed action sequences.
(Summers-Stay et al., 2012) use an action grammar and extract an activity tree from human motion as well as object contact relations. The activity tree is built based on events in the action grammar. Events can consist of bringing two objects together, resulting in a new object, or using a tool together with a single object, resulting in a transformed object. The objects in the activity tree originate from such events, such as *a piece of bread on a plate*, from bringing together a slice of bread and a plate, which results in a tree structure. Each event triggers adding new children to a tree leaf node. With this deep hierarchical representation it is possible to segment and represent long action sequences. Actions used for evaluation in this paper are typical household actions such as making a sandwich.
(Ramirez-Amaro et al., 2014) proposed a similar approach in the same scenario. They learn a decision tree from hand-object relations like *hand moving towards an object* or *object in hand*. Whenever these rules lead to a state change, a new key frame is inserted. This approach is similar to the semantic segmentation presented in this thesis, yet the rules are more specific since they are also used for action recognition.
(Pardowitz et al., 2008) propose a method for segmentation based on gestalt laws and a Competitive Layer Model with hand and object trajectories as input data. (Hendrich et al., 2010) and (Barchunova et al., 2011) present approaches for segmentation of manipulation skills based on multi-modal sensor input. (Barchunova et al., 2011) use data gloves, a contact microphone and contact-force sensors as input for a Bayesian inference method (Fearnhead, 2006) to segment a manipulation sequence of a single object (since the microphone is attached to the object). They are able to segment actions like holding, (un)screwing a bottle or shaking. A similar sensor modality is used by (Matsuo et al., 2009) to segment in-hand manipulation tasks into manipulation primitives. They equipp an object with force-sensor skin and measure the contacts of the human hand on the object. From the force measurements,

they create a *contact force index*, which is a generalized representation of all contact forces and torques on the object. The *contact force indices* are clustered with the expectation maximization algorithm to assign a group of *contact force indices* to manipulation primitives. Whenever the current *contact force index* cluster changes a new primitive starts and thus, a new segment begins.

3.1.4 Discussion

Most segmentation approaches focus on the extraction of segments from motion data for specific applications or on the extraction of general activities like walking, running, jumping etc. One drawback of such approaches is their lack of generalization capability to high variance motions, which often contain targeted motions, such as grasping. Additionally, data-driven approaches or template-matching methods are fixed to the training data that is used, which might be suitable if the target application has a fixed set of actions, but is unsuitable if general motions have to be segmented. No approach focuses on segmentation of manipulation actions since their motions are not distinctive enough solely based on motion data. Approaches based on semantic information tackle this problem by using motion independent features such as object relations. Yet semantic approaches only succeed if semantic changes are observed, which is not always possible for actions such as tossing a bottle that do not cause any observable state changes. To this end, the approach presented in this thesis leverages both feature spaces, motion data and object relations, to achieve a robust segmentation of different types of manipulation actions.

3.2 Robot Skill Programming and Execution

Modeling and programming of robot skills require expert knowledge and development experience. The task becomes even more difficult if complex robots like humanoid robots with a high number of sensors and actors (degree of freedom, DOF) are the target platform. This can be driven into the extreme if robot-agnostic robot skills are to be developed. Modeling such skills in general purpose programming languages is difficult due to the requirement to coordinate several sensor and actor systems, which all work independently from the host at their own frequency. Therefore, modeling methods that target the requirements of robot skills are needed. Many papers have dealt with this challenge and most papers propose hierarchical and graphical modeling such as statecharts.

3.2.1 Statechart Concept and Variations

Besides the original publications (Harel, 1987; Harel and Politi, 1998), a multitude of further publications (Samek, 2002; Coleman et al., 1992; Von der Beeck, 1994) and software projects (Angermann et al., 2014; Yakindu, 2015; EasyCODE, 2015) on statecharts for a variety of use cases exist. In this thesis, the focus regarding statecharts lies on software realizations of statecharts and on statecharts in the context of robotics. However, not all semantics of the original statecharts are seen as feasible by authors of derived approaches. To this end, most papers change or extend the semantics of statecharts to their needs and opinion.

(Von der Beeck, 1994) and (Breen, 2004) critically discuss several features of the statechart formalism such as orthogonality and history states. Breen advocates that orthogonality should often be solved by independent statecharts and that history states can weaken the concept.

There are several projects offering frameworks for developing your own statecharts. In late 2015, the W3 consortium released version 1.0 of an XML statechart notation (ScXML, (World Wide Web Consortium (W3), 2015)) to establish one format describing Harel statecharts. Similar, the Object Management Group defined the UML StateMachines notations (Object Management Group (OMG), 2015). Yet both mainly specify general purpose notations of the Harel formalism. The approach and realization presented in this thesis is a ready-to-use statechart framework designed for use in robotics. In UML it is also possible to specify control and data flow graphically in an activity diagram. The equivalent of states are actions, which are connected via activity edges called *ControlFlow* or *ObjectFlow*. *ControlFlow* edges are only capable of activating their target actions, while *ObjectFlow* connects two *ObjectPins* and can carry data values. This feature is similar to the data flow of the statechart concept presented in this thesis, yet the *ObjectPins* only provide very rudimentary data flow features such as carrying one data field between two actions.

The well-known de-facto extension of C++ Boost (Huber, 2007) contains a subproject called the Boost Statechart Library, which offers a statechart implementation close to the original formalism of Harel. It has the unique feature to specify statecharts with C++ templates and to achieve compile-time statechart validation. While this is a valuable feature to ensure valid statecharts, it does not fit the requirements of experimental robotics, i.e. quick adjustment of parameters and the statechart structure for short development cycles. On the side of graphical tools, the graphical statechart modeling tool QM (Quantum Leaps, 2015) provides means for designing and implementing event-driven low-level statecharts for embedded systems with a strong focus on traceability at the code level. The statechart structure is generated into C++ code, i.e. each state and transition is represented by its own C++ class or function. This means that code regeneration and recompilation are necessary for every statechart structure change. In statecharts presented in this thesis, the goal is to generate code only to catch errors in the user code as early as possible and for auto-completion purposes in Integrated Development Environments[1] (IDE), but not for functional features.

In (Yakindu, 2015) another graphical statechart modeling tool is presented, aiming at usability and assistance inside the editor during typing. Never-

[1] Integrated Development Environments (IDE) are powerful text editors with additional functionality like compiling, debugging and convenience features for editing.

theless, it seems to target low-level statecharts like QM with limited data flow control. Data is stored in global variables and is accessible from any state as it is done in many other approaches. Yet data flow control is of high importance in the statecharts presented in this thesis as described later.
Statecharts are used in robotics to control behavior on a high-level in several approaches (Klotzbücher and Bruyninckx, 2012; Merz et al., 2006; Bohren et al., 2011; Billington et al., 2010; Thomas et al., 2013) since statecharts address several of the problems in robotics like state-based control and event-triggered execution. Most of the approaches change or extend the original formalism to their identified requirements of robotics. Several approaches remove some functionality since the original statecharts are too unrestrictive for reusability and composition of complex robot skills (e.g. (Klotzbücher and Bruyninckx, 2012)). Others add more functionality, mostly for less complex statecharts (e.g. *inner transitions* in (Angermann et al., 2014)). (Stampfer and Schlegel, 2014) present an aspect similar to the dynamic structure of statecharts presented in this thesis. They replace states with alternatives from a "robot app store" to increase robustness and reduce complexity.

restricted Finite State Machine (rFSM)

In many aspects, the statecharts presented in this thesis are similar to the restricted Finite State Machine (rFSM), presented in (Klotzbücher and Bruyninckx, 2012), from the Orocos framework (Bruyninckx et al., 2003). They also try to address the shortcomings of the original statechart formalism of Harel. rFSMs are a minimal subset of UML2 and of Harel statecharts consisting of the three elements *states*, *transitions* and *connectors*. *Connectors* are a unifying representation for the *initial*, *junction*, *entry* and *exit* pseudostates of UML2.1 for a reduced formalism. Similar to (Angermann et al., 2014), one addition is *internal transitions* for complexity reduction in some use cases. Concurrency is omitted in rFSMs since they claim that the way how concurrency can be implemented is unclear and that the feature is not needed in robotics since coordination as done in rFSM is always a short action. Distribution of a statechart over several hosts is also discussed. Yet they introduce a container state which adds two more state hierarchy levels just for using a distributed state instead of the completely transparent approach as proposed in this thesis.

Overall, rFSMs focus on coordination of components, but offer only very limited support to specify transition based data flow. Figure 3.5 shows an example of an rFSM for coordinating a gripper with entry and exit actions and guarded transitions. The authors promote the "pure coordination" concept, where the coordination part of the framework should be strictly decoupled from the computation capabilities to avoid unresponsiveness and blocking. This resembles the state phases of the approach presented in this thesis. The state phases are split into coordination and computation phases. To give the developer the ability to easily create critical sections, separation is only encouraged and not enforced in the here presented statechart framework.

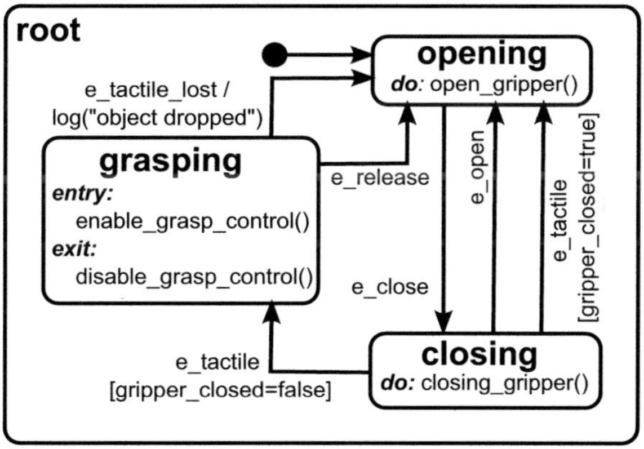

Figure 3.5: Example for coordinating a gripper with restricted Finite State Machines (rFSM) for modeling hierarchical robot skills. *Source:* (Klotzbücher and Bruyninckx, 2012) © Joser 2012

SMACH

In the well-known robot operating system ROS (Quigley et al., 2009), an approach called SMACH (Bohren et al., 2011) is employed that focuses on data flow in statecharts. However, the scope of data flow in ROS SMACH is handled differently in comparison to the data flow in the statechart approach presented in this thesis. In ROS SMACH, a child state can access all data used by its parent state. This eases programming because it is easy to use data on several levels, but also violates the principle of modularity of states and creates implicit data dependencies between states. A state using data fields of a parent state cannot easily be reused in another state since it depends on the availability of specific data fields in a parent state. Due to this, data scopes over several state levels are not allowed in the statechart concept proposed in this thesis and explicit mapping of data between state levels is required. Furthermore, ROS SMACH only supports graphical online visualization of states, but does not provide any tools for graphical programming.

Simulink: Stateflow

Stateflow (Angermann et al., 2014) is a highly integrated statechart approach of the software toolkit MatLab, which allows convenient design of behaviors. In combination with the simulation environment Simulink, their statechart approach can easily be connected to a robotics environment and be utilized for developing robot skills and behaviors. In Stateflow, not all features of the original statecharts are incorporated, e.g. global events are not possible. They also introduce new features such as *temporal logic* for execution based on the passed time or *condition actions*, which are actions connected to events instead of transitions and are executed if the event occurs. Another feature is *inner transitions* which are similar to reflexive transitions, but skip the exiting and entering action of a state, which can greatly simplify the structure of a statechart. But the biggest selling point of Stateflow remains the tight integration into the powerful frameworks of MathWorks to quickly create running systems.

3.2.2 Graphical Programming Tools in Robotic Frameworks

When developing high-level software on a robotic platform, it is desirable to configure and connect existing components using a graphical user interface to prevent writing repetitive and therefore error prone source code. This allows new as well as experienced users to intuitively and efficiently combine mid- and high-level components in order to create a functional system structure. Since writing software is one of the main challenges in robotics for beginners, such as students, Graphical Robot Programming offers a great entry point. It removes the obstacle presented by syntax and control flow of a conventional programming language (Rahul et al., 2014). Graphical software development often combines complexity reduction by connecting modular components on a macroscopic scale with the option to write low-level software (e.g. joint controller) (Pot et al., 2009). Graphical and tabular representations are an accessible way to model system behavior in the context of simulation, validation and consistency checking of a system design before final implementation (MathWorks, 2015c). (Hirzinger and Bauml, 2006) are using Simulink (MathWorks, 2015b) in conjunction with MATLAB (MathWorks, 2015a) to graphically model subsystems and generate executables running on a realtime target. The Microsoft Visual Programming Language (Microsoft, 2012b), as part of Microsoft Robotics Developer Studio (Microsoft, 2012a), proposes developing the complete logic and program flow in a visual development environment as it lowers the bar for novice programmers. However, in this thesis, visual development is deliberately limited to the definition of structure, of used data types and of data flow in the statecharts for the benefit that the user can write unrestricted C++ code. The RDE YARP (Metta et al., 2006) also offers means of graphical programming with the *gyarpbuilder* (Paikan, 2014), yet on another level. With *gyarpbuilder* it is possible to connect continuous input and output data of components graphically and to insert arbitrators in these connections to manipulate data flow easily. *RtcLink* (AIST, 2017) from the OpenRTM project offers a GUI to operate on real-time components in a network. It can activate and deactivate components as well as connect their ports. It leverages the capabilities of an established IDE by providing the GUI as an Eclipse plugin.

3.2.3 Other Robot Skill Modeling Approaches

Besides approaches based on statecharts, like the approach presented in this thesis, many other representations have been utilized or developed to model and program robot skills. Some approaches rely on Finite State Machines (e.g. (Aein et al., 2013; Loetzsch et al., 2006)). One of the first papers presenting a graphical robot programming system is (Naylor et al., 1987), but due to the early years of robotics and computers the focus of this paper lies on graphical editing and less on graphical modeling.
(Loetzsch et al., 2006) present the Extensible Agent Behavior Specification Language (XABSL) for describing behaviors of autonomous agents based on hierarchical finite state machines. This approach was used by teams competing in the RoboCup (Lötzsch et al., 2003) to design the behavior of soccer robots. XABSL is a simple programming language and a behavior consists of four components: Agents, options, states and decision trees. An agent is here the top-level entity and consists of a number of behaviors, which are called options. Options are hierarchical compositions of sub-behaviors to form complex behaviors from primitive behaviors. Each option is a finite state machine with states that define actions of the agent. These actions can reference other options allowing for hierarchical decomposition of a task. Transitions in the finite state machines are triggered based on a decision tree. These decision trees utilize sensor data or input data in combination with basic arithmetic and Boolean logic to form decisions.
(Asfour, 2003) and later (Lehmann et al., 2006) propose to use Petri nets for parallel execution of tasks with error detection and exception handling. The parallelism of Petri nets is used to simultaneously execute tasks of different

subsystems, such as a mobile platform and an arm. An additional Petri net is used for supervision of the execution to allow for interruption by the user or a planning system. (Ziparo and Iocchi, 2006) introduce Petri Net Plans as a representation of multi-agent robot plans and actions. The petri nets are composed of several basic constructs such as *ordinary-action*, *sensing-action*, *conditionals*, *loops* and more. Furthermore, they extended petri nets with labeled/conditional transitions for coordination with external input or triggers by the operator.

(Fraser et al., 2016) present an approach for a fixed, hierarchical, compact robotic task representation that allows for easy specification of a task process. Based on three sub-task types (AND, OR, THEN), task constraints can be specified for unordered and ordered execution and dynamic action selection based on the current world state. Which sequence of action is chosen follows a top-down decision process which is fed bottom-up by confidence values of potential actions about the applicability of the action. Though, as intuitive and simple as the representation is (which is an important aspect in robotic task specification) it seems very limited in regard to flexible, sophisticated skills. The task representation cannot be parameterized to new situations that are not covered during task representation design, and there is no data flow between nodes. Thus, the adaptability is restricted to the situations specifically modelled in the design phase. Failure handling also seems to be marginal and might lead to starvation and/or invalid paths: If one action node of an AND/THEN-node never reaches the activation threshold (but the average of all nodes is high), the system will starve although there might be a valid path somewhere in another branch of the task representation.

Behavior-based systems (e.g. (Arkin, 1998; Nicolescu and Matarić, 2002; Frank et al., 2012; Paikan et al., 2014a,b)) are another way to specify high-level robot functionality. The most striking difference is that statecharts are state-based and behavior-based systems are rule-based. This means state-charts have an explicit current state, while behavior-based systems only have an implicit state. Additionally, behavior-based systems are inherently parallel whereas statecharts are sequential. While behavior-based systems may be closer to behavior of humans or animals, their scalability for programming complex capabilities is questionable. For a human, an explicit state is easier to comprehend. Moreover, it eases the debugging process. Both are vital criteria for software development and maintenance.

3.2.4 Discussion

This section has given an overview over the state of the art regarding robot skill modeling and programming. While many robotic research groups still use unstructured plain code with long if/else or switch statements, a multitude of approaches for structured modeling of robot skills is available. Most of these use graphical modeling approaches such as statecharts for better comprehensibility. Others use Domain Specific Languages to simplify and specialize the programming language on the target domain. While several graphical approaches (e.g. (Klotzbücher and Bruyninckx, 2012; Angermann et al., 2014; Bohren et al., 2011)) are comparable to the statechart approach presented in this thesis, they all fail to consider the data flow in robot skills adequately. Complex robot skills such as visually guided grasping coordinate a multitude of different algorithms such as object recognition and localization, motion planning and motor controllers, which each require a set of parameters to function correctly. To achieve composable, reusable and adaptable or even robot-agnostic robot skills some of these parameters need to be adjusted for different tasks or different robots. Some approaches allow for limited data flow specification (Angermann et al., 2014; Bohren et al., 2011) for single data values and/or at a large scope, but it is necessary to be able to specify in detail the interface of a skill, i.e. the input and output parameters, to achieve truly reusable skills. (Bohren et al., 2011) keep all variables at the statechart scope. While this seems practical, it violates the reusability principle and opens the door for unwanted side effects during repeated execution of a statechart.

Furthermore, robot skill modeling approaches for complex systems like humanoids need to provide means to cope with distributed systems and failures of sensors and subsystems. Since humanoid robots are thought to perform tasks like humans, tasks are likewise complex. This means that the representation needs to be able to represent low-level actions such as platform navigation and high-level capabilities like symbolic planning operators (e.g. *grasp object X*). The statechart approach presented in this thesis tackles these problems as described in chapter 5 and provides a clearly structured and comprehensible representation.

3.3 Execution of Complex Tasks on Humanoid Robots

Single robot skills are useful to achieve one specific goal. But combining them to dynamic action sequences or plans increase the possibilities greatly. Yet how skills can be successfully combined requires advanced knowledge or complex reasoning about the current state of the world: object locations differ, object relations change and the robot might be at a different position. All these changes require different parameterization of the robot skills or even different action sequences.

One prerequisite for executing tasks in a dynamic environment is to perceive and understand the world. In other words, the current state of the world needs to be considered when executing a task.

There are two fundamentally different types of approaches: Approaches that model tasks directly and approaches that solve tasks based on some reasoner and a knowledge base. A literature review reveals many different approaches on this topic. Researchers started working on this topic several decades ago with e.g. a theoretical hierarchical planning and execution system as demonstrated in (Nilsson, 1973). The idea is to use short plans, which recursively consisted of more detailed plans themselves. Nilsson also considered failures and surprises during execution, which is a crucial part of any task execution system, by backtracking the full plan to the problematic action and solving the task again from that point. This process is propagated up the hierarchy until a level can find a new solution. Yet Nilsson pointed out that porting such a system onto a real robot would require substantial additions.

Similarly, Hierarchical Task Networks (HTN) by (Erol et al., 1994) try to divide-and-conquer the planning problem by specifying actions and tasks, of which the tasks themselves consist of actions and tasks. This approach combines using predefined tasks and reaching a goal with a planner. The authors compare their tasks to "recipes" for a problem. Both types, actions and tasks, can have preconditions, whereas only actions can have direct effects since tasks are always just compositions of actions. From logic point of view, each task contains or just provides a set of suitable action/task candidates. In this way, it is possible to reduce the complexity by magnitudes. Though, additional knowledge and often more knowledge engineering by the developer is required for each task. Another advantage is the possibility to plan on state variables instead of predicates. The main difference to classical planning (see section 2.5) is not to achieve a goal, but to perform a task. The search algorithm itself in two implementations of the authors, PyHop[2] and SHOP[3], is a simple search, where each task path is followed recursively until a subtask has failed. In this case, the plan path is backtracked to the last decision point and the next option is chosen.

In contrast to this, classical planners like FastForward[4] (Hoffmann and Nebel, 2001) or Planning with Knowledge and Sensing (Petrick and Bacchus, 2002), only apply planning operators, i.e. actions, to the world state to achieve a given goal. Yet less knowledge engineering is needed here to create a valid domain.

The Task Description Language (TDL) introduced by (Simmons and Apfelbaum, 1998) is an approach which uses designed task descriptions and gen-

[2] `https://bitbucket.org/dananau/pyhop`

[3] `http://www.cs.umd.edu/projects/shop/`

[4] `https://fai.cs.uni-saarland.de/hoffmann/ff.html`

erates trees as representation of the robot tasks. These trees consist of two different types of nodes: *command* nodes and *goal* nodes. *Command* nodes specify behavior or robot skills that are executed upon entering the node. *Command* nodes can only have other *command* nodes as children. *Goal* nodes represent high-level behaviors and are expanded at run-time into other *goal* nodes or *command* nodes. Without further constraints all children of a node are executed concurrently. But since robot actions often need to be executed in a particular order it is possible to specify additional constraints for each node such as passing of time or that another sibling node has to be finished first. Similarly, termination constraints can be specified, which terminate the execution of a node prematurely if a specific event occurs. Semantically, there are many parallels between TDL trees and statecharts. Yet TDL trees are by default concurrent whereas statecharts are by default sequential. Figure 3.6 shows an example of a TDL tree.

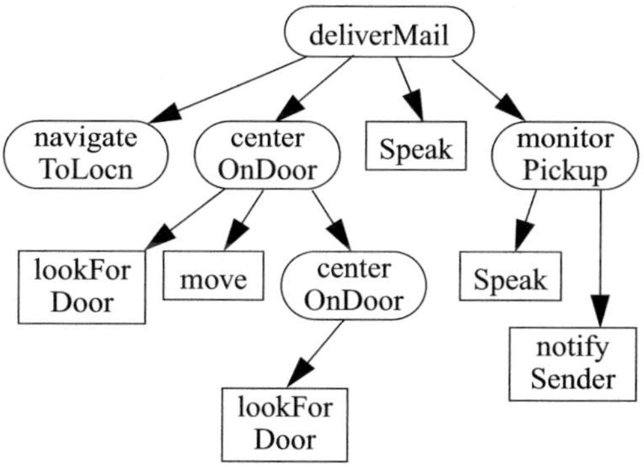

Figure 3.6: Example of a Task Description Language tree describing one task. *Source:* (Simmons and Apfelbaum, 1998) © 1998 IEEE

The shift of focus in the recent years goes towards integrating such task description or solving systems into full robot systems with sophisticated natural language understanding (Ovchinnikova et al., 2015; Dzifcak et al., 2009; Nyga and Beetz, 2012a; Matuszek et al., 2013; Liu and Zhang, 2016),

reasoning about the environment (Agostini et al., 2015; Beetz et al., 2011) and execution monitoring (Aein et al., 2017; Konecny et al., 2014).
An approach focusing on execution monitoring based on planning is presented by (Konecny et al., 2014). Their method considers not only symbolic effects of actions during monitoring, but also temporal aspects. They show that even failures of actions without observable effects can be detected by using causal, temporal and categorical knowledge.
(Agostini et al., 2015) propose a method for planning and plan execution in dynamic environments with the perk of object replacement based on affordances. The system is evaluated in a cooking task. First, a plan is generated from a prototypical problem definition. This plan is checked for unavailable objects in the current scene. If any object is missing, a reasoning engine in combination with a database (Szedmak et al., 2014) is used to find objects with matching affordances. If a replacement is found, the missing object in the plan is replaced and the execution starts without actual replanning. Action FSMs, which were previously learned from demonstration, are used for execution (Aein et al., 2013). The success of an action is measured by evaluating the effects of actions based on the changes of the object-relations. The replacement method is similar to the one proposed in this thesis, yet the proposed one incorporates more modalities. Another difference is that they plan first and replace afterwards whereas in the proposed method the planning follows the replacement process. This way, the costly planning process is only triggered if known objects are available. Additionally, locations are proposed by the replacement method. (Boteanu et al., 2016) also present an approach for object replacement leveraging big semantic networks like WordNet[5] (Pedersen et al., 2004).
(Beetz et al., 2011) propose another system for preparing a meal. Here, the cooking task is performed by two robots with a different range of duty. The

[5] `http://wordnet.princeton.edu`

recipes are downloaded from the World Wide Web, e.g. from a *how-to*[6] website for humans. They create a sketchy-plan from the natural language recipe in technical terms. Each step of a recipe is represented as a goal to be achieved. A symbolic planner uses this to generate an action sequence to achieve this goal. The robot also utilizes ontologies to determine probable places of objects. For example, the pancake mix is flagged with the attribute *perishable*. *Perishable* objects are stored in the fridge. Therefore, the robot generates the hypothesis that the pancake mix might be in the fridge, which is needed for the plan generation. This feature is similar to the location hypothesis generation proposed in this thesis. Yet different knowledge bases are used. Whereas (Beetz et al., 2011) use an object ontology based on object attributes, the approach proposed in this thesis uses exchangeable, multi-modal strategies to generate hypotheses.

One important part of an autonomous robot acting in an unstructured environment is the capability to process and store the perceived knowledge for future usage. (Beetz et al., 2015) present a robot knowledge system called openEASE[7] that strives to develop an online knowledge base for robots that can be accessed and filled by robots all over the world to extend the reasoning capabilities and profit from the learned capabilities of other robots. The knowledge base includes semantic information as well as an episodic memory about the world state and trajectories of the agents. To provide structured access to the information a complex query language is employed. This knowledge base is similar to the robot memory system *MemoryX* that is used in this thesis.

(Bollini et al., 2013) also had the vision to create a kitchen chef robot called *BakeBot*, which can understand recipes written for humans and which creates a feasible plan and also executes this plan. This work is similar to the work of (Beetz et al., 2011). The recipes are also downloaded automatically from the

[6] `http://www.wikihow.com`
[7] `https://www.open-ease.org`

Internet and parsed into known, parametrized baking action sequences. It is possible to extract multiple action sequences from one recipe text. Therefore, the extracted baking action sequence is optimized based on a reward function, which is learned from a labeled dataset of recipes with their correct action sequences. The FastForward planner (Hoffmann and Nebel, 2001) is used to generate motion primitive sequences from the baking action sequences.

(Lisca et al., 2015) propose a system using *Probabilistic Action Cores* (PRAC), see (Nyga and Beetz, 2012b, 2015), that obtains task descriptions from natural language instructions, generates a plan based on the task description. The plan is then executed with predefined control programs. This system is evaluated in the context of tasks in a biological laboratory and in particular in a DNA extracting procedure on the PR2 robot.

3.3.1 Discussion

(Nilsson, 1973) pointed out that the execution of plans must be well integrated into the generation of plans with means to compare the expected results with the actual execution results. This is one important aspect that has not received appropriate attention up until today. Most planning frameworks only offer a file based interface for calculating the full plan and do not take the monitoring of plan execution into account. This is probably due to the fact that all benchmarks assess how fast planners create full plans in a simplified simulated world. Evaluation on real robots is rarely performed due to the required high effort.

Whether to use classical planning or HTNs for a sophisticated service robot system depends on the application. Classical planners cannot deal, unlike HTNs, with domains containing a multitude of constants and actions. Yet HTNs require more knowledge engineering since not only the actions, but also higher level tasks need to be modeled. Another decision factor is whether it is feasible or more suitable to generate goal descriptions (classical planning) or task descriptions (HTNs) for the robot.

Although a lot of progress has been made in the area of reasoning and planning for service robots in dynamic environments, all approaches are far away from a general solution to the problem due to the complexity of the whole systems and each of the subsystems. Those systems need to integrate several subsystems such as language understanding, perception, reasoning and execution. Since not even the problems of the subsystems can be seen as solved also the compositing systems are incomplete. Therefore, all approaches target specific applications such as cooking a meal. The approach presented in this thesis improves several aspects of the existing systems such as symbol replacement and on-the-fly generation of a complex symbolic planning domain completely from the robot's memory state. One of the main concepts of the execution system is the exchangeability and extensibility. While several replacement strategies (see subsection 6.3.2) are already presented in this thesis, new modalities can be added even during runtime. Further, the domain translation (see section 6.5) can easily be adapted to new environments and different features of a robot.

3.4 End-to-End Task Learning from Demonstration Approaches

While there are many papers dealing with subproblems of learning from demonstration (LfD) many researchers have also presented approaches dealing with all aspects needed for a functional system, i.e. end-to-end approaches. These approaches need to provide solutions for the understanding of the demonstration, the representation of the learned information and for the execution. Here, approaches focusing on task learning from demonstration are presented. A general survey about learning from demonstration can be found in (Argall et al., 2009).

LfD is composed of three steps: observation, representation and execution. Although every step is necessary for a complete LfD system, some approaches skip or simplify parts of the process, e.g. (Mohseni-Kabir et al.,

2015) demonstrate in simulation or (Dillmann, 2004) assumes that primitive actions are already available. This is due to the fact that LfD is a vast research area with a series of difficult subproblems, which constitute a reasonable research problem themselves, e.g. the segmentation of demonstrations as elaborated in section 3.1.

As a first step, the capturing of the demonstration needs to be processed. Here, some approaches convert the demonstration in a generalized representation (Jäkel et al., 2010) or use a representation that is tightly linked with the next step (Aksoy et al., 2011; Ramirez-Amaro et al., 2014). Most approaches segment and classify the demonstrations as the next step (Kuniyoshi et al., 1994; Dillmann, 2004; Aksoy et al., 2011; Ramirez-Amaro et al., 2015) before generalizing (Jäkel et al., 2012; Mohseni-Kabir et al., 2015; Ekvall and Kragic, 2006, 2008) and transforming the demonstrated task into an often hierarchical representation (Dillmann, 2004; Zöllner et al., 2005; Mohseni-Kabir et al., 2015). In the various approaches, the execution phases are different since they need to be suitable for the employed robot. Symbolic planners managing primitive actions are utilized in this phase (e.g. (Mohseni-Kabir et al., 2015; Nicolescu and Mataric, 2003)), or own representations are tailored to their requirements (e.g. (Chang and Kulić, 2013b)).

(Kuniyoshi et al., 1994) present one of the first approaches for an end-to-end system for learning from observation. They use a multi-camera system to capture the human demonstration by tracking the hand and the objects. Segmentation and classification based on motion types, manipulated objects and action effects are utilized to understand the demonstration, which is used for bottom-up inference of the demonstrated plan. For execution, the stored plans are instantiated on a 6 DOF robotic arm with a gripper and adjusted to the currently perceived environment. Whereas this sounds like everything is already solved, every component (certainly due to hardware limitations) was rudimentary compared to today's systems.

(Nicolescu and Mataric, 2003) present a system based on behavior networks. Their system is able to generalize observed sequences into common action sequences over multiple demonstrations and extracts temporal dependencies as well as preconditions needed for each action. Furthermore, the human can provide feedback to the robot to correct or improve tasks.

A programming by demonstration system is presented by (Dillmann, 2004) in which a training center consisting of hand trackers, data gloves, multiple cameras and a microphone is used for capturing the demonstrations. An abstract, system invariant task knowledge representation builds the base for mapping demonstrations from a human to a robot. Tasks are demonstrated via action sequences, which are pruned of irrelevant actions during a generalization step. The actions are organized in a tree structure based on a rule-based system. The actions themselves are assumed to be available and not described in the paper. (Zöllner et al., 2005) presented a similar approach with the additional benefit of alternative actions.

In (Ekvall and Kragic, 2006) an approach was presented that is able to learn from multiple human demonstrations by decomposing each task into subtasks for segmentation and classification. Furthermore, the segmented tasks are inserted into a task model, which describes their goals and constraints. For online execution a task level planner is employed that enables the robot to choose the best strategy depending on the current environment.

(Jäkel et al., 2012) presented an approach for learning manipulation tasks from demonstration by utilizing constraints observed during demonstration and constrained motion planning. The constraints consist of force, contact and collision constraints. The initial constraints are extracted from the demonstration of the teacher and are then adapted to the kinematics of the robot. Generalization is done based on different demonstrations, which have to cover well the possible variations of the task, e.g. large distances between two goal positions of a placing action would lead to a gap in between those demonstrations in the generalized model although it might be possible to place the object there. This problem is alleviated by additionally testing more configurations in simulation. For perception of the demonstrated task, multiple cameras, datagloves, motion trackers and tactile sensors are used.

The system is evaluated on two complex and difficult bi-manual manipulation tasks on a real two armed robot: Unscrewing a bottle while holding it and lifting a spoon with one hand to grasp it with the second hand.
An approach based on Petri nets is proposed in (Chang and Kulić, 2013b) and an extension in (Chang and Kulić, 2013a). The petri nets are automatically created from one or multiple demonstrations and used to generate action sequences for imitation of an observed task. The latter paper extends the approach by an error recovery mechanism.
(Wörgötter et al., 2015) present a concept called structural bootstrapping for improving robot skills on all abstraction levels through exploration and accommodation and assimilation of new knowledge. Structural bootstrapping is a probabilistic process inspired by child language acquisition that combines existing knowledge with new observations to supplement missing information to the robot's knowledge base about planning-, object- and action-relevant entities. They use means of task segmentation and action semantics learning (Semantic Event Chains (SEC) - (Aksoy et al., 2011)), affordance learning (Xiong et al., 2013), symbolic planning (Petrick and Bacchus, 2002) and plan recognition (Geib, 2009) to infer unknown actions and objects in observed plans und thus learning new plans from observation.
Two approaches based on Hierarchical Task Networks (HTN) are presented by (Mohseni-Kabir et al., 2015) and (Boteanu et al., 2016). (Mohseni-Kabir et al., 2015) focus on an interactive learning process between the human teacher and the robot, in which the robot can ask for help and additional information based on collaborative discourse theory. The robot is able to learn a task from a single demonstration or multiple demonstrations, which are merged into one HTN based on the work presented in (Mohseni-Kabir et al., 2014).
(Ramirez-Amaro et al., 2015) transfer skills to a humanoid robot based on semantic rules and a knowledge base. The authors claim to extract the "essence" of an activity, which means which aspects of the demonstrations are needed to achieve the goal. The semantic rules ground on object relation

information such as which tool or object is involved in an action. Since semantic information is utilized, the approach is invariant to different timings or execution styles and one scenario can be transferred to new situations. The execution on the robots makes use of a pre-existing primitive library which is parametrized according to the observed action.
A library of actions with semantics is learned in (Aein et al., 2017) based on the observation method and the action semantics proposed in (Aksoy et al., 2011). Each semantic state change rule is associated with a set of primitive actions that in combination achieve this change. Tasks are represented by linking them to a specific set of these rules. A FSM is used for execution of tasks, which selects appropriate actions and monitors the fulfillment of action effects. For evaluation a large set of different objects is used in 30 trials for each of ten tasks and for two more complex, composited tasks (object rearrangement and making a salad).

3.4.1 Discussion

Although many papers have been published in the area of learning from demonstration (LfD) there exists no complete solution that can solve every task and application and it will take several more years or decades until LfD is solved. On the contrary, most approaches work only for a very limited range of applications and only under certain conditions. E.g. the approach of (Ramirez-Amaro et al., 2015) only distinguishes between three motion types (*move, not move, tool use*) and object relations, which do not differentiate between a grasp and a pushing action. In most approaches, demonstrations are carefully adapted to the system regarding motion style and speed since natural motions are more difficult to perceive and segment. Therefore, the segmentation approach in this thesis is targeted to improve segmentation results on natural, seamless motion. While action recognition works well, most limitations arise when it comes to the execution on robots because then detailed, robot adapted representations are necessary. The difficulties

lie here in the several modalities that are relevant for different actions, a high precision and the correspondence problem between human and robot. Therefore, a powerful hierarchical skill programming approach is used in this thesis for representing skills of all modalities.
Additionally, it is a challenge to reason on the world and adapt robot skills to it. To tackle this problem, this thesis presents an approach that can plan on incomplete environment knowledge by proposing hypotheses for objects and locations and utilizing these to plan goal fulfilling action sequences.
Table 3.1 shows a comparison of the discussed approaches based on several criteria: The criterion *observation* describes what sensors are used. This determines what kind of information is available and how precise the data is. The criterion *situation adaption* describes if an approach can adapt to situations that are different from the demonstrated situation. A more advanced adaption is described by the *online planning* criterion. This describes if a task can be extended or restructured to solve the task at hand. *Skill learning* refers to whether the underlying skills or actions are learned from demonstration as well or if predefined skills are used. The *skill/task representation* states, if available, the name of the used representation for execution of skills and tasks. The last column, *symbol replacement*, describes whether the approaches can suggest alternatives for symbols, e.g. objects and locations, that lead to an acceptable solution.

Table 3.1: Comparison of task learning from demonstration approaches.

	Observation	Situation Adaption	Online Planning	Skill Learning	Skill/Task Repres.	Symbol Replacement
(Kuniyoshi et al., 1994)	Gray-scale Video	Yes	Yes	No	-	No
(Nicolescu and Mataric, 2003)	RGB-Video, Speech	No	Yes	No	-	No
(Dillmann, 2004)	Training-Center	Yes	No	No	-	No
(Jäkel et al., 2012)	Training-Center	Yes	Yes	Yes	Strategy Graphs	No
(Chang and Kulić, 2013b)	RGB-Video	Minimal	No	Yes	Petri-Nets, DMP	No
(Ramirez-Amaro et al., 2015)	RGB-Video + AR Marker	No	No	No	-	No
(Aein et al., 2017)	RGB-D	Yes	No	No	SEC, DMP	No
(Mohseni-Kabir et al., 2015)	Simulation	Yes	Yes	No	HTN	No
Presented Approach	Marker-based Motion Capture	Yes	Yes	No	Statecharts	Yes

4 Task Understanding by Hierarchical Segmentation and Action Recognition

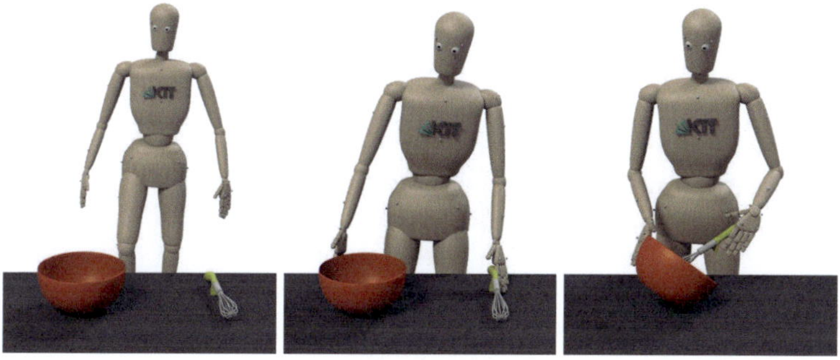

Understanding natural human demonstrations is an indispensable ability of robots that should learn from observation. Yet understanding demonstrations is a difficult endeavor. The number of possible demonstrations in an environment with several objects is not manageable if each demonstration is seen as one element. Alongside with the divide-and-conquer paradigm, a first step towards understanding is segmenting a demonstration into meaningful segments to reduce the complexity of the observation. Fortunately, complex tasks like preparing a meal consists usually sequences of actions, such as *grasp, mix, place etc.* However, a human performs these actions in a seamless stream without clear transitions between these actions, which increases the difficulty to segment such demonstrations into a sequence of actions.

Most task segmentation algorithms focus on poses or motion trajectories of the human (see subsection 3.1.2), whereas a few algorithms focus on semantic features extracted from the world state. Nevertheless, both feature spaces alone are not distinctive enough to extract all segments: Motion based methods have difficulties to generate a correct segmentation if one action follows another seamlessly, e.g. if an object is first grasped and then pushed, there might not be any significant change in the motion between these actions. Based on motion trajectories it is also difficult for humans to perform such a segmentation. In general, it is likely that two persons segment the same demonstration differently (Lin et al., 2016b).
Semantic approaches face other problems. They depend on the fact that the state of the world changes. However, if the detection of these changes fails, no segmentation point can be extracted. For example, detecting the effect of shaking a fluid in a bottle with current state of the art observation systems is difficult. Fortunately, in many situations problematic action transitions in one feature space are easy to deal with in the other feature space.
Yet the segmentation of a demonstration does not reveal which actions it consists of. The segments are just unlabeled motions and world states, which need to be associated with known descriptions of actions. The is known in the literature as action recognition. In most approaches, action recognition is similar to segmentation approaches performed based on the motion data (Aggarwal and Ryoo, 2011) and in some cases based on semantic features (Aksoy et al., 2016). But some actions, especially goal-directed manipulation actions, do not have a similar motion trajectory. For example, the motion of a grasping action from a table in contrast to the motion of a grasping action out of a box. The motion trajectory for the object on the table is a fairly straight motion whereas a curved motion is needed for grasping out of a box. Semantic features, however, dot not always change during actions, which makes recognition of such action with semantic approaches impossible.
In this thesis, a hierarchical task segmentation approach which uses semantic features in combination with motion based features is proposed. This chapter

describes an extended version of the approach presented in (Wächter and Asfour, 2015).

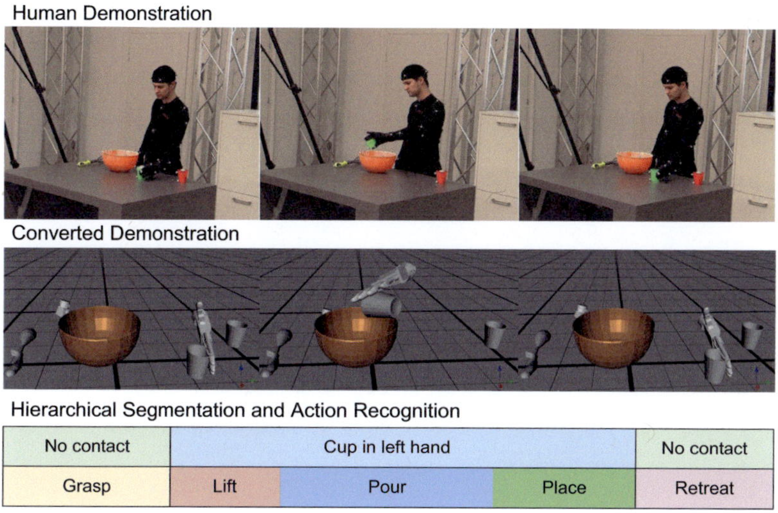

Figure 4.1: Overview over the segmentation and recognition approach.

The hierarchy consists of two levels (see Figure 4.1):

- the semantic level (section 4.3)
- the motion characteristic level (section 4.4)

The features of the semantic level are extracted from object contact relations. The segmentation is premised on the assumption that every change in the semantic feature space must be caused by an action of the demonstrator, whereas the features of the motion characteristic level describe the dynamics of a motion. The top-level of the hierarchy uses the semantic features since these features are more reliable than motion based features. Thus, the motion based features are used on the bottom-level of the hierarchy for sub-segmentation of the semantic segments and are used to find segments

missed by the top-level. In the next two sections, the data representation that is needed to apply this algorithm will be explained.
A novel approach for action recognition is presented in section 4.5, which leverages similarly to the segmentation approach both feature spaces, motion and semantic feature space, to create a meaningful action descriptor that is used for learning a decision tree.

4.1 Representation of Human Demonstrations

In order to capture human demonstrations with a high accuracy and at a high resolution, the marker-based motion capture system VICON[1] is employed. In contrast to most other approaches, the objects and the environment are captured in addition to the human. To be able to capture all entities, markers have been attached to the human subject as well as to objects of interest present in the current scene (see Figure 4.2).

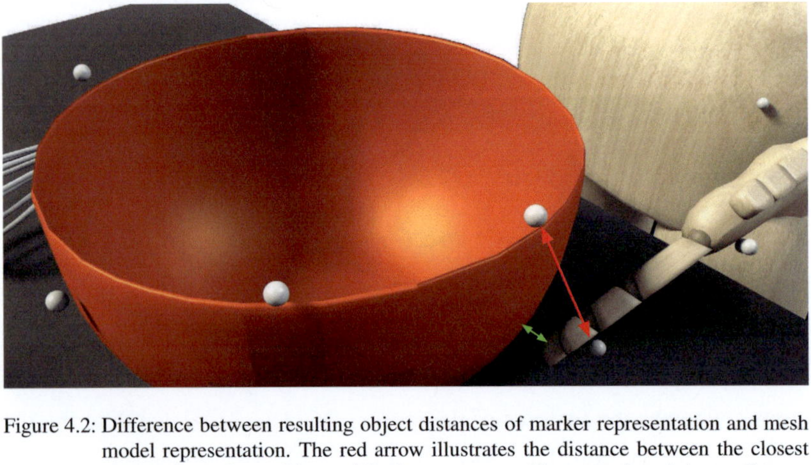

Figure 4.2: Difference between resulting object distances of marker representation and mesh model representation. The red arrow illustrates the distance between the closest two markers of two objects while the green arrow illustrates the minimum distance between the mesh models. The mesh model distance is significantly smaller.

[1] http://www.vicon.com

Since tracking the whole body of the human does not improve the segmentation, only the hands of the human are considered during the post-processing. The hands are treated as rigid bodies as the fingers are not tracked. Three markers on each object and each hand are sufficient to calculate the 6D pose of the object or hand. Yet more markers increase the robustness in case of occlusions. All markers are labeled and grouped according to the object they belong to. The motion capture data consists of 3D Cartesian space trajectories of all markers. However, the trajectories and the label of the markers do not sufficiently describe the shape of the object for the proposed segmentation approach. Figure 4.2 illustrates the difference between representation consisting only of markers and the representation with 3D mesh models. To enhance the recordings, the trajectories are converted from a marker representation to a 6D object pose representation with motion converters of the Master Motor Map framework (Terlemez et al., 2014). The converters require a mapping of the labeled markers from the recordings to virtual markers on a 3D mesh model. Thus, 3D mesh models of each object were created with a 3D object scanner (Kasper et al., 2012) or common 3D modeling tools and equipped with virtual markers on the model. The resulting object representation consists of a 3D mesh model of each object as well as the positions of all labeled markers attached to this object. This information allows to calculate the 6D object pose using the 3D shape registration algorithm described in (Besl and McKay, 1992). In Figure 4.3, the mapping of observed and virtual markers is visualized.

To retrieve the 6D pose trajectory for an object, the transformation is calculated for each frame and applied to the base pose of the object. This conversion is performed for all captured objects, which leads to an accurate representation of the scene during the demonstration. In the applications presented in this thesis, trajectories with at least three markers for each object in any frame were used. If a marker is missing for a short time in the recordings, the marker position is interpolated based on the other markers belonging to its marker group.

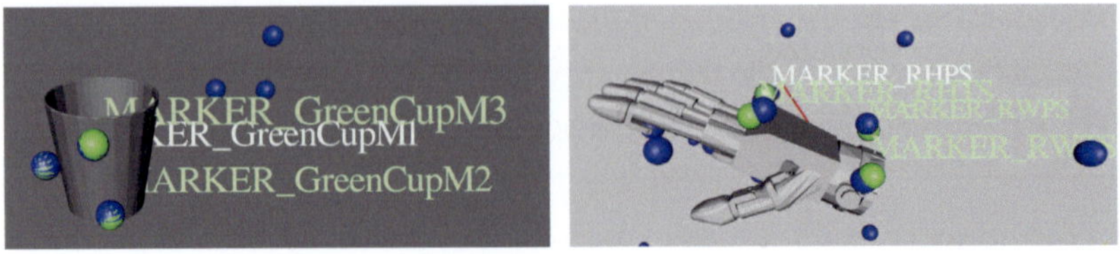

Figure 4.3: Mapping of observed markers (blue spheres) attached to an object to virtual markers (green spheres) on its model. Both, observed and virtual markers, are aligned with the 3D shape registration algorithm of (Besl and McKay, 1992). Text in the figure describes the labels used for identification of the markers. *Left:* The visualization of the cup's 3D mesh shows the 6D pose of the perceived object. *Right:* For non-rigid objects, like a human hand, the relative positions of the markers change depending on the deformation of the object. Since the object model is static, this results in larger errors of the marker mapping. These errors are in the case of a hand small enough to not affect the segmentation algorithm.

4.2 Input Data for Task Segmentation

As input data for the hierarchical task segmentation the motion data of the human as well as of the objects and the environment are used. In the following, these are referred to as *entities*. A similar approach was proposed by Aksoy et al., which uses an RGB stereo camera (Aksoy et al., 2010, 2011) or RGB-D (Aksoy et al., 2015) images as input. While this approach does not require to model each involved object, it does not provide 6D trajectories of human hands or objects and lacks robustness in terms of action execution velocity and precision. The authors estimate contacts between objects by recognizing overlapping color blobs. In this work, marker-based motion capture is used to record demonstrations. As mentioned earlier, the shape and the pose of the entities are not sufficiently represented by the markers alone. A segmentation based on the distances between the entities requires the use of high distance thresholds to detect contact points that are far from the markers. This reduces the robustness of such a method since entities in the demonstrations need to have a relatively large minimum distance between each other. The introduction of a 3D mesh model as entity representation instead of markers allows the use of sophisticated mesh-based collision detection algorithms, such as the one described in (Larsen et al., 1999), to accurately calculate the distance between objects and to detect contacts or collisions between them.

A demonstration D is represented by the 6D trajectories of all entities and the distances between all entities for each time frame:

$$D = ((M_1, \cdots, M_e), (F(1), \cdots, F(T))) \tag{4.1}$$

$$F(f) = ((\vec{p}_1(f), \cdots, \vec{p}_e(f)), (d_{1,2}(f), \cdots, d_{1,e}(f) \ldots d_{e-1,e}(f))) \tag{4.2}$$

where $F(f)$ defines the pose of each entity and the distances between all entities at frame f, T is the number of frames, e is the number of entities in the demonstration, $d_{i,j}(f)$ is the distance between the entities i and j at frame f, $\vec{p}_i(f)$ is a 6D pose of entity i at frame f and M_i is the mesh model of entity i.

Figure 4.4 shows all the intermediate data representations during the segmentation process.

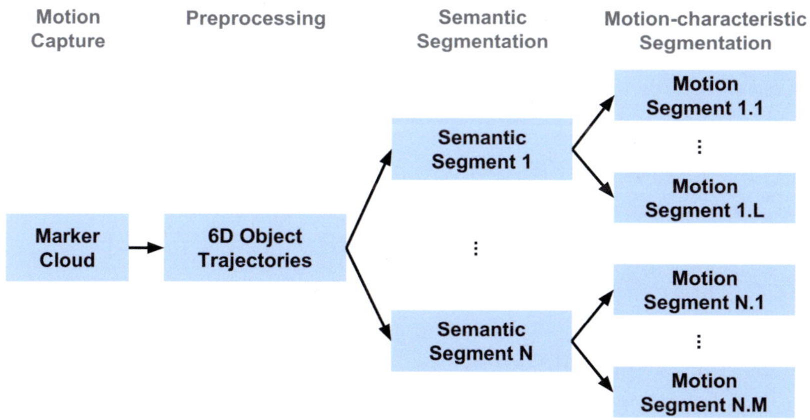

Figure 4.4: Several data representations are used in the process of the hierarchical segmentation as illustrated here.

4.3 Semantic Segmentation based on Object Relation Changes

On the top level of the hierarchical segmentation, the human demonstration is segmented into semantic segments based on semantic changes in the world state similar to the idea presented in (Aksoy et al., 2010). Semantic segments are motions which have an observable semantic effect on the world state or as (Ziaeefard and Bergevin, 2015) wrote "What does it mean to do an action?". This world state consists of entity contact relations, which are extracted based on the 3D mesh model and the 6D pose of each object. Thus, the demonstration is segmented by detecting key frames based on the change of relations between entities. Only contact relations are considered, where $contact(A,B)$ denotes contact between entities A and B. Other relations like *on* or *in* ground on such contact relations and do not improve the results of the segmentation method. Yet they might prove valuable for action recognition. The relation $contact(A,B)$ relies on the closest distance between any part of the involved entities A and B. $contact(A,B)$ returns true if the distance falls below a predefined threshold ε. To deal with noise in the distance measure, hysteresis is applied on the threshold if a contact has been detected in the last frame.

It is possible that the first frame with contact between two objects is not the moment that is considered by a human as the end/start of a motion. For example, if the human grasps an object, the fingers might touch the grasped object while still approaching the object. The final grasp pose, i.e. the key frame, might occur later when the grasp is stable. To account for this, key frames are extracted at the local minimum of the distance curve between two objects. Figure 4.5 illustrates this behavior. Additionally, to compensate motion capture inaccuracies the relative velocity between two objects needs to be similar, i.e. the objects need to move in the same direction with similar velocity.

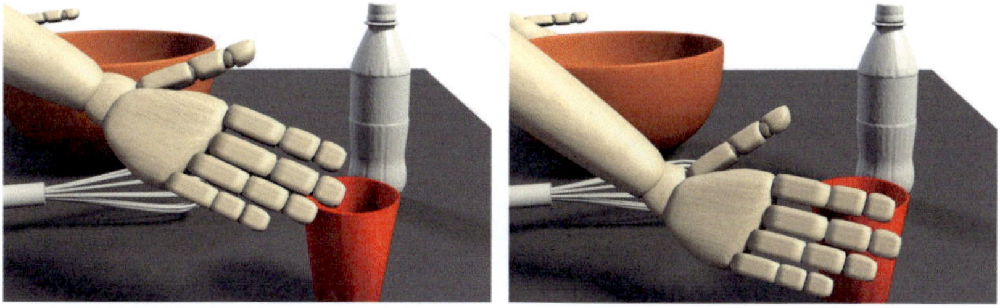

Figure 4.5: When approaching an object the first contact (left) between two entities is not always the desired key frame. Therefore, the key frame detection is deferred until the contact relation is stable (right).

The three relations between objects are:

1. *absent*-Relation:

$$absent_f(A,B) = !A_f \vee !B_f, \tag{4.3}$$

   where $!A_f$ or $!B_f$ means that the object A respectively B is not present in frame f.

2. *contact*-Relation:

$$contact_f(A,B) = \begin{cases} \begin{matrix} (d_{A,B}(f) < \varepsilon) \\ \wedge \quad l_{A,B}(f) \end{matrix}, & !contact_{f-1}(A,B) \\ \begin{matrix} (d_{A,B}(f) < \varepsilon * \lambda) \\ \wedge \quad (|\hat{\vec{v}}_{A,B}(f)| < \tau) \end{matrix}, & contact_{f-1}(A,B), \end{cases} \tag{4.4}$$

   where $d_{A,B}(f)$ is the Euclidean distance between entity A and B, λ denotes the hysteresis factor. τ is a velocity difference threshold below which the velocity of two objects is considered similar. $l_{A,B}(f)$ is true if a frame is a local distance minimum between object A and B:

$$l_{A,B}(f) = \begin{cases} true, & (d_{A,B}(f) < d_{A,B}(f-1)) \\ & \wedge \ (d_{A,B}(f) < d_{A,B}(f+1)) \\ false, & otherwise \end{cases} \tag{4.5}$$

$\hat{\vec{v}}_{A,B}(f)$ is the Cartesian velocity difference between entity A and B:

$$\hat{\vec{v}}_{A,B}(f) = \vec{p}\,'_A(f) - \vec{p}\,'_B(f) \tag{4.6}$$

3. *no_contact*-Relation:

$$no\_contact_f(A,B) \ = !contact_f(A,B) \tag{4.7}$$

All three relations are mutually exclusive.
Calculating these relations for all frames and all entity combinations results in the semantic representation R of the demonstration as a three-dimensional array $R^{T \times e \times e}$ with:

$$R_{ijk} = \{absent, contact, no\_contact\} \tag{4.8}$$

Whenever the relation between two entities in two consecutive frames changes its status, a frame is considered a key frame k and added to the set of key frames K. This results in the segmentation subarray S:

$$S^{T \times e \times e} = (s_{kij})_{k \in K, i, j \in \{1 \dots e\}} \tag{4.9}$$

This matrix encodes the semantic world state as relations between all objects for each extracted key frame. Figure 4.6 depicts several snapshots of a drinking demonstration and the corresponding object relation matrices. Figure 4.7

shows an example of a demonstration sequence including actions such as grasping, shaking, tossing and pouring to which this segmentation algorithm was applied.

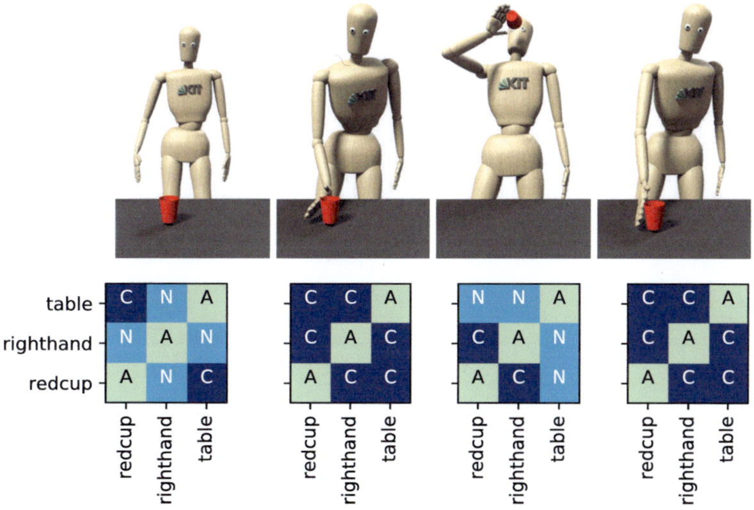

Figure 4.6: Visualization of multiple semantic world states during a demonstration. The matrices describe the relations between the objects and the corresponding state seen in the 3D visualization. The abbreviations mean: A: Absent, N: No Contact, C: Contact. The colors show the same information as the letters. From the human model only the hands are considered for the segmentation.

4.3.1 Post-processing: Merging of Key Frames

Not all actions correspond to only one relation change. For example, the action of *dropping* a ball (B) from a hand (H) into a container (C) can be associated with two relation changes (under the assumption that liquid can be tracked):

$$\begin{matrix} contact(B,H) \;\wedge \\ !contact(B,C) \end{matrix} \rightarrow \begin{matrix} !contact(B,H) \;\wedge \\ contact(B,C) \end{matrix}$$

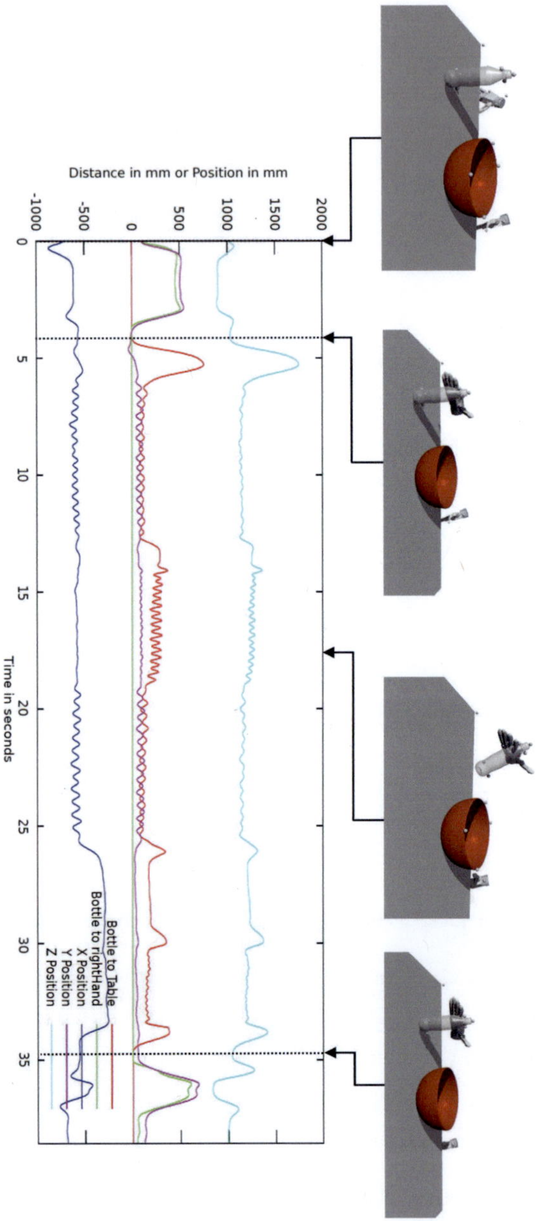

Figure 4.7: Segmentation based on object contact relations. Whenever the contact relation between two objects changes, a new key frame is extracted (black dotted vertical lines). The pictures above show the corresponding frame of the demonstration.

Hence, key frames need to be merged into groups of key frames that semantically belong together. For most actions, these key frames appear within a small time window as it is the case for the action of *dropping* an object into another. A simple way to cope with this is to consider the temporal displacement of two key frames and to merge them if this distance between the key frames is too small.
State changes are always instantaneous, although e.g. *pouring* might seem as a continuous state change. However, the change of the contact relation does not take time. If the *pouring* takes noticeably long, it would result in two key frames: A first key frame when the liquid gets in contact with the target container and a second key frame when the liquid loses contact with the source container. Thus, the first would be the starting key frame and the second key frame would be the end of the *pouring* action.

4.4 Segmentation based on Motion Characteristic

The first level of segmentation described above results in a segmentation of the human demonstrations in semantic segments that have observable changes in the world state. Nevertheless, some actions have unobservable effects, even for a human. For example, the effect of shaking two transparent liquids in a bottle cannot be observed visually. In this and other examples, such unobservable effects are relevant for segmentation and ultimately for understanding the demonstrated task. However, their detection based on current state of the art methods and sensor technologies is challenging.
The previously described method can only detect events when two entities come in contact with each other. Yet, it is desirable to also detect actions without observable effects. To this end, the segmentation from the previous section is extended with a sub-segmentation that extracts motion parts within a semantic segment based on the trajectory shape and the motion characteristics. In other words, the detected semantic segments serve as the input for the

bottom-level of the hierarchy and are divided further by the motion characteristic segmentation. The goal of the sub-segmentation is to decompose the semantic segments into smaller parts that contain motions with different motion characteristics and potentially represent the different motion primitive within a semantic segment. Many motion based segmentation algorithms use insufficient features for extracting atomic actions since they target longer activities like walking or dancing (Barbic et al., 2004; Zhou et al., 2013) or cannot handle high variances in action repetition (Lin et al., 2014). For example the Zero-Velocity-Crossings algorithm in (Fod et al., 2002) detects key frames when the motion changes the direction or stops completely. Thus, actions with seamless transitions cannot be detected and false positives occur frequently for periodic actions. Segmentation based on Principal Component Analysis or Dynamic Movement Primitives extracts segments with an upper complexity limit of the motion. But a complexity limit does not describe a distinctive feature for an action: A grasping motion is often almost linear while a shaking motion has a complex trajectory profile.

Yet several motion based segmentation methods could be used for this sub-segmentation. In this work, a novel heuristic is used that assesses the characteristic of a motion and divides a motion when the characteristic changes. To capture the characteristic of a motion, the approach uses the dynamics of the motion as a basis, i.e. the acceleration values of the trajectory. Figure 4.8 continues the segmentation shown in Figure 4.7 and shows an example for this motion segmentation applied to the same demonstration sequence.

There are two fundamentally different ways to segment motion data. One type is to find key frames that meet a specific criteria, and the other one is to search for meaningful segments. The presented approach lies in between. The approach searches for key frames that maximize the difference of the trajectory parts left and right of a key frame. As such, the approach differs from the pure key frame search since the key frame itself is unimportant. It also differs from segment search because it does not require the complete

demonstration segments to be known in advance. In short, the proposed approach segments the trajectory in most distinctive parts.
To find the key frames, the demonstration trajectory is analyzed recursively. On every recursion level, the given trajectory segment is searched sequentially with a predefined step size for the key frame that divides the trajectory best. Subsequently, the segments left and right of this key frame candidate are analyzed again in the same manner until the segment size falls below a threshold or no additional segments with a sufficiently good quality can be found. The whole approach is described in Algorithm 1.

Algorithm 1 Motion Characteristic Segmentation Algorithm

```
function FINDKEYFRAMES(kf, t_l, t_r, l_min, s, μ)
    // kf: in-out parameter; intially empty key frame list
    // t_l,t_r: timestamps of current segment borders
    // l_min: minimum segment length
    // s: stepsize for sliding window
    // μ: minimum segment quality
    for t := t_l + l_min to t_r − l_min ; t += s do
        for d := 0 to dimensions do
            q_n ← CALCQUALITY(t, d)
            if q_best < q_n then
                q_best ← q_n
                t_best ← t
            end if
        end for
    end for
    if q_best > μ then
        kf.INSERT(t_best,q_best)
        FINDKEYFRAMES(kf, t_l, t_best, l_min, s, μ)
        FINDKEYFRAMES(kf, t_best, t_r, l_min, s, μ)
    end if
end function
```

To define the quality of a frame, which is needed to decide whether a frame is a key frame, first the following terms are introduced:

$$s_{l,d}(t_c) = \int_{t_c-\frac{w}{2}}^{t_c-1} \sqrt{1+a_d'(t)^2}dt \left(\frac{\hat{U}_l}{\hat{U}_r}\right)^2 \tag{4.10}$$

$$s_{r,d}(t_c) = \int_{t_c}^{t_c+\frac{w}{2}-1} \sqrt{1+a_d'(t)^2}dt \left(\frac{\hat{U}_r}{\hat{U}_l}\right)^2, \tag{4.11}$$

where $a_d'(t)$ is the derivation of the acceleration vector of dimension d at timestamp t, d is the dimension of the trajectory, $s_{l,d}(t_c)$ and $s_{r,d}(t_c)$ are the quality scores left and right respectively of the key frame candidate, t_c is the timestamp of the key frame candidate, w is the window size left and right of the key frame candidate that is analyzed. $\hat{U}_l$ and $\hat{U}_r$ are the peak-to-peak amplitudes of the acceleration left and right of the key frame candidate. Equation 4.10 calculates the score of the segment left of the key frame by calculating basically the length of the function. Equation 4.11 does the same for the right side of the key frame. To also consider the amplitude of the acceleration, the score is multiplied with the squared relation of the peak-to-peak distances left and right of the key frame candidates. Finally, the quality q_d of a key frame candidate is then defined as:

$$q_d = \begin{cases} \frac{s_{l,d}}{s_{r,d}}, & s_{l,d} > s_{r,d} \\ \\ \frac{s_{r,d}}{s_{l,d}}, & s_{l,d} \leq s_{r,d} \end{cases} \tag{4.12}$$

So far, the qualities for each dimension are normalized by their amplitudes. However, the amplitude of one dimension can be small compared to another dimension. Since motions in a dimension with overall low amplitudes are

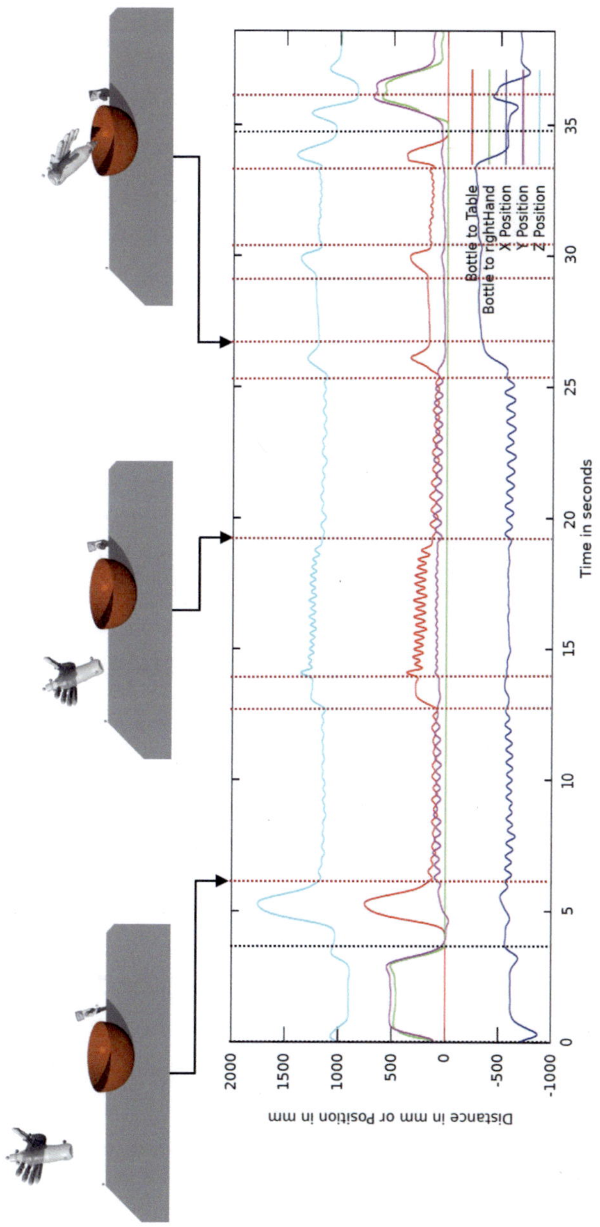

Figure 4.8: Segmentation based on the change of the motion characteristic. Recursively the frame with the biggest change in the motion characteristic is searched. The found key frames are shown as red vertical dotted lines. Black vertical dotted lines are the borders of semantic segments.

not as important as another dimension with high amplitudes, the qualities for each dimension are aligned with the maximal peak-to-peak distance $\hat{U}_d$ of all dimensions:

$$\hat{q}_d = q_d \cdot \sqrt[z]{\frac{\hat{U}_d}{\max_d \hat{U}_d}}, \tag{4.13}$$

where z is a scalar which influences the weight of the normalization. The key frame with the best $\hat{q}_d$ of all frames and dimensions is selected as a key frame with the quality q, if the value does not violate a quality-threshold or a minimum segment size to avoid oversegmentation.
The idea of this heuristic is that motions with a different characteristic, e.g. smooth circles, intense shaking, pouring have a different acceleration profile and therefore a different shape. The heuristic primarily measures the length of the acceleration curve and normalizes it with the amplitude of the segment.

4.5 Action Recognition

The segmentation presented in the previous sections is able to segment complex manipulation demonstrations into meaningful segments, i.e. actions. Yet the segmentation does not provide a grounded symbolic meaning of the segments. It only provides semantic states before and after the actions. In the literature, the process of classifying each segment is called *action recognition*, in which each segment is labeled with a known symbolic identifier such as *grasp* or *pour*. Similar to the segmentation approach, the action recognition approach presented in this section uses semantic as well as motion features. Most action recognition approaches rely solely on motion features such as poselets (as presented in the survey by (Ziaeefard and Bergevin, 2015)) and a few rely on semantic features (e.g. (Aksoy et al., 2016)). Yet some actions do not provide enough distinctive information in each of the feature spaces. For example, a *grasp* action depends heavily on the location of the grasped object and the motion trajectory differs therefore greatly between repetitions of the action. Such an action is difficult to recognize with motion based features, but exhibits an obvious state change in the semantic feature space: The contact relation between the hand and the target object changes. The action *shaking* a bottle is difficult to recognize in the semantic space since the mixing of the contents of the bottle cannot be observed by technical means. Yet this action shows promising features in the motion trajectory space with a rapid, periodic motion of the end-effector. Thus, using both feature spaces should increase the action recognition rate. But there are even more benefits from combining both feature spaces than recognizing actions which only show features in one of the feature spaces. For example the actions *wiping* and *stirring* are very similar in the motion feature space, but in combination with the semantic states this ambiguity is easily solved since *wiping* actions are executed with different tools than *stirring* actions. To this end, both feature spaces, semantic and motion features, are used in this approach and combined into one feature descriptor.

Semantic Features for Recognition

The semantic recognition features utilize the same information as used for the semantic segmentation: the object contact relations. For each segment, i.e. action, the semantic state of the first frame is taken as the first part of the semantic feature descriptor. The possible object-relations are *absent*, *no_contact*, *contact*. The state at the last frame of the segment is not used directly, but used to calculate the state delta between the first and the last frame of the segment. The state delta is the second part of the feature descriptor. Since most classification algorithms work on float arrays, the semantic feature descriptors need to be also represented as such. Additionally, the dimensionality of the training data for all demonstrations needs to be the same. A suitable representation for this is a two-dimensional distance matrix as depicted in Figure 4.9. Because the object-relations are symbolic, these relations need to be converted into a float or integer representation. To be able to calculate a state delta from two semantic state matrices with the same distance between each possible relation type one-hot encoding is used. This translates

- *absent* to $(1,0,0)$,
- *no_contact* to $(0,1,0)$ and
- *contact* to $(0,0,1)$.

Thus, each matrix cell contains a three-dimensional vector. To keep the size of the matrix constant between all demonstrations each entity used in any of the demonstrations is encoded into the matrix. Additionally, an object hierarchy is used to generalize actions, e.g. all manipulated objects are also represented by the row/column of the matrix labeled *object*. Since multiple objects influence the state of the *object* relations, the relation types are ordered and the highest occurring relation type is used. The order of the relations is *contact* > *no_contact* > *absent*. Objects that are not present during a demonstration are flagged as *absent*. Combining the distance matrix with the one-hot encoding creates a three dimensional matrix, which is rolled out into a one dimensional vector for the classification algorithm. The state delta is created by subtracting the matrix of the semantic state after the action from the semantic state matrix before the action. This means if 10 objects are used, the feature vector for the semantic features is $10^2 * 3 * 2$ floats long, where the square results from the distance matrix, the 3 from the one-hot encoded cells and the factor 2 from the state matrix and the state delta matrix.
Figure 4.9 shows an example segment. The visualizations at the top depict the demonstration, in which the human approaches the red cup with the right hand. The matrices below show the object contact relations at the start of the action, at the end of the action and the state delta.
Reflexive relations are always marked as *absent* since they cannot appear. The *absent* relation is also important for objects that are not present during one demonstration, but are used in other demonstrations since all objects appear in the full relation matrix, as explained later. In this example, the red

cup touches the table in the beginning and the table and the hand at the end of the action (C entries, meaning contact, in the matrices). The hand also touches the table in this demonstration, but it might not do so in other demonstrations. These variations between demonstration repetitions are considered later in the classifier.

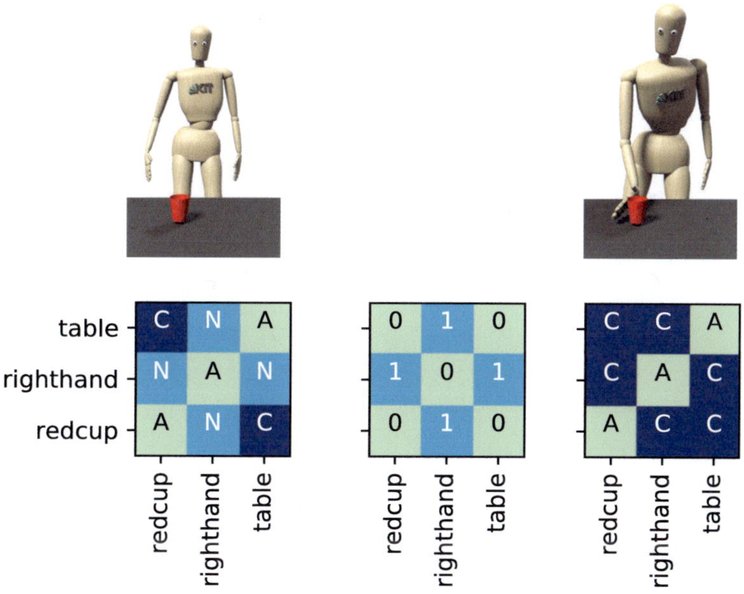

(a) Action start (b) State delta (c) Action end

Figure 4.9: Example of semantic states of an *approach* action. (a) and (c) show the semantic state before and after the action. (b) shows the state delta between (a) and (c). (a) and (b) are used as the semantic features of the action recognition. The abbreviations mean: A: Absent, N: No Contact, C: Contact. The numbers in the delta matrix show the simplified difference of the states.

Motion based Features for Recognition

The semantic states of an action already reduces the search space of possible actions so that only a few actions need to be distinguished. For example, when holding a bottle the demonstrator will probably not try to write or cut.

Thus, the motion based features used here do not need to be as distinctive as they need to be in a general motion based action recognition approach. Furthermore, it should be possible that actions with largely different motion trajectories can be in the same action class, e.g. the *approach* action can have very different motion trajectories depending on the location of the object. The motion based recognition features should, in line with the motion based segmentation heuristic, consider global characteristics of a motion such as the dominant frequency.
The following global motion based features are used for the action recognition:

- **Dominant frequency**: The dominant frequency is extracted for each of the dimensions of the Cartesian space. Since calculating the dominant frequency with the Fourier transformation proved to be not reliable, a geometric approach was chosen that calculates the average distance between local minima and maxima of the position. This feature is important to detect periodic motions and is only used if a minimum number of cycles is detected, otherwise the value is set to zero.

- **Intensity:** The intensity characterizes how dynamic and fast an action was demonstrated based on the mean of the absolute acceleration values.

- **Main direction:** This feature represents the main direction of the action, which is typically towards the human or away from him.

- **Work space:** This feature is calculated based on the amplitude of the motion trajectory and represents the extent of the motion.

Each of the features is calculated for each Cartesian dimension, resulting in a twelve dimensional float vector for each action.

Decision Trees for Action Recognition

The feature vectors are merged and used to train a general purpose classifier. As the feature vector contains mostly irrelevant dimensions for a single action and the training only covers a very small part of the feature space, a classifier is required that can handle sparse data and detect relevant dimensions. Most classifiers do not work on such data because they need dense data, but decision trees are well suited for this problem. Here, classification and regression trees (CART) are used (Loh, 2011). Support Vector Machines, Bayesian Networks, Gaussian Processes and Neural Networks were also evaluated, but produced significantly worse results.

Each action is one training data element represented as a float vector of size $(12 + \#objects^2 * 3 * 2)$ and multiple labels. The labels describe different abstraction levels, e.g. grasping a bottle is labeled as *graspBottle* and *grasp*, to be able to generalize the observed action, e.g. to abstract from the used objects. To train the classifier as many repetitions of actions as possible need to be fed to the classifier. To account for variations in demonstrations, such as irrelevant contacts of an action, multiple demonstrations should be available in the dataset. The decision tree classifier will then be able to detect the relevant dimensions.

The prediction on the learned model uses the same data format and can recognize actions in an online manner.

4.5.1 Accommodation as new Object-Action Complexes

Planning operators are symbolic descriptions of action preconditions and the action effects on the world state, which are used by planning systems to manipulate a world state in order to achieve a given goal. The preconditions of a planning operator define a subset of the world state predicates with unbound variables that need to be fulfilled to apply this planning operator. The effects are the changes, which the planning operator has on the world state and to which the world state of the segmentation result is similar. Each of the extracted segments represents elementary actions, of which only a limited number of different actions is needed. According to (Wörgötter et al., 2013) most manipulation tasks can be composed of only 27 actions. The accompanying world states at the key frames before and after a segment correspond to the world state before and after a planning operator was applied. The label of the recognized segment can be used to find the matching planning operator. But also the parameters of the planning operator need to be found. This can be done by binding the parameter variables of this planning operator to the objects of the demonstration by using the object hierarchy and comparing the observed effects with the effects of the planning operator as proposed in (Wächter et al., 2013).

Extracting entirely new operators from the observation by machine learning is a research area of its own and was already addressed by others (e.g. (Mourao, 2012)) and is therefore omitted in this thesis. New composite planning operators representing a full demonstrated task can be learned by analysis of the contained basic planning operators. Each occurring object instance represents one parameter of the composite operator. The preconditions are calculated by adding up all preconditions of the contained operators if they are not fulfilled by any previous operator effect. The effects are calculated by adding up all effects of the contained operators and by deleting predicates that occur with the same amount of positive and negative instances. The preconditions and effects are to be stored in the Object-Action Complex (see chapter 2) segment of the long-term memory together with the list of actions that are parametrized with the parameters of the new composite planning operators. The new planning operator increases the planning performance because it executes several actions in one step and can contain actions without observable effects, which a planning system would never use since it cannot see their effect. Since planning domains, especially if they are generated automatically, can explode easily in complexity, i.e. if the branching-factor during each planning step is high, an observed task stored as one planning operator can make the difference between solvable and unsolvable tasks.

4.5.2 Object Hierarchy

The objects used in the demonstration and in the planning operators are all known and organized in an object class hierarchy since planning operators can often be applied to more than one object class. Only leaves of the hierarchy tree can appear in the demonstration or in the binding of a planning operator. Every element of the hierarchy can have multiple parents and multiple children. Having multiple parents of one object class allows specifying heterogeneous sets of objects. E.g. a cup can be *graspable* and *pourable* while a hammer is also *graspable*, but not *pourable*. Thus, if the *grasping* planning operator contains a parameter of type *graspable* the cup and the hammer are candidates while for the pouring action with parameter type *pourable* only the cup will be considered.

4.6 Summary

In this chapter, a hierarchical segmentation approach and an action recognition approach based on semantic information using object contact relations and motion features were presented. The novelty lies in the consequent usage of both feature spaces and in exploiting the orthogonality of the feature spaces. The hierarchical segmentation approach consists of two levels, the semantic level and the motion characteristic level. The semantic segmentation has precedence over the motion characteristic level since the semantic features based on object contact relation changes are more grounded. The motion based segmentation extracts segments based on significant changes in the motion characteristic. Yet the point at which a change is significant cannot be pin-pointed onto one value and can depend on the types of the demonstrated actions.

The action recognition uses both feature spaces to reduce the ambiguity in each individual feature space for certain actions by leveraging the orthogonality of the feature spaces. Actions that cannot be distinguished in one feature space have often clearly different representations in the other feature space. These recognized actions correspond to robot skills that can be modeled with the approach presented in the next chapter.

5 Distributed Statecharts for Hierarchical Robot Programming

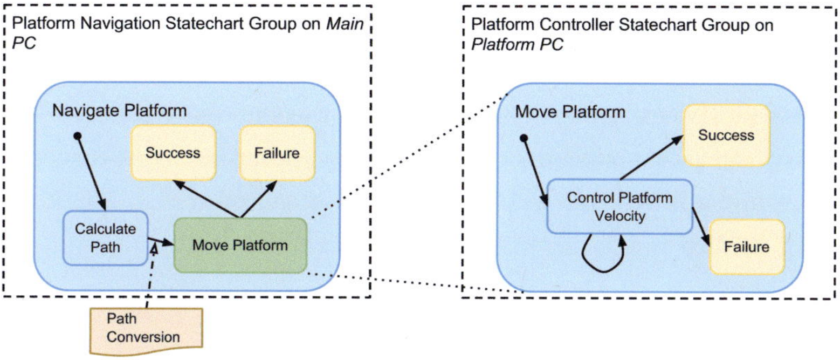

In the previous chapter, an approach for segmentation of a human demonstration into a sequence of actions has been presented. Yet from observing actions it is not possible to just replay these actions successfully on a humanoid robot. The embodiment and uncertainties in perception as well as the state of the environment are too different to just imitate these actions. Therefore, a representation of actions or even complete tasks is needed to enable a robot to reproduce observed tasks.

Dealing with the complexity of multi-component systems can be challenging in terms of control and data flow. Hence, only skilled experts are capable of designing and realizing highly connected software systems as they are needed for humanoid robots. In this work, the statechart concept is proposed to provide a representation for robot behavior and skills that reduces complexity while increasing reusability of already created functionality.

The statechart concept is part of the robot software development environment ArmarX (Vahrenkamp et al., 2015). A robot framework in ArmarX consists of several distributed components, which provide access to sensors and actors (i.e. the hardware), offer computational functionality and implement a robot memory system as a common data source for the robot software. On top of these robot components, the statechart mechanism simplifies the definition of the coordination on all abstraction levels of robot behavior (i.e. the program flow). By separating coordination from behavior, the task of building new robot software applications can be supported through graphical user interfaces while maintaining full flexibility on source code level. In order to gain full flexibility within the robot application, the programmer can use well-defined entry points to implement user-specific source code. In the following, the design principles chosen for the proposed statechart concept and the resulting differences to Harel's formalism are presented. This chapter was published in a shorter version in (Wächter et al., 2016).

5.1 Design Principles

Key design principles of the proposed statechart approach are: modularity, reusability, runtime reconfigurability, decentralization and state-disclosure.

- *Modularity* of the proposed statechart concept is realized through the individual states and their explicitly specified input and output. There is no direct interaction allowed between substates of different parent states. Such direct transitions would violate the defined interface of a state.

- *Reusability* is ensured, since every state can be used as a substate in any other state and has a specific interface for interaction. The interface is specified with state parameters similar to parameters of a function.

- *Runtime-reconfigurability* means that a statechart can be defined in configuration files and every aspect of the structure can be changed at runtime.
- *Decentralization* means that a statechart does not need to reside in one process, but can be spread over several processes and hosts. This enables load balancing and increases robustness by enabling crash recovery (see subsection 7.2.1).
- *State-disclosure* means that the current state and all its parameters can be inspected at runtime and logged for future behavior adaptation.

5.2 Differences to Harel Statecharts

The proposed statechart concept differs in several points from Harel's original formalism. Some of Harel's features are omitted to comply with the stated design principles and to simplify the statechart design process for the developer. One important aspect was added to the statechart concept, which is not covered in Harel's formalism: data flow specification and control during transitions. The *condition*-connectors and *hierarchies* are available like in the original statecharts. Direct *inter-level-transitions* are not allowed in order to avoid violations of the principle of modularity. The *history*-connector of Harel's formalism is not available since it conflicts with the data flow specifications. Furthermore, the *history*-connector is removed to reduce side-effects during execution as well as to simplify the comprehension of the current state of the system during introspection. Each entering of a state with the same parameters must provide the same internal state. *Orthogonality* is currently only available in a smaller scope. Each active state can contain an asynchronous user code function executed in a separate thread. Thus, the different hierarchy levels can run in parallel.

5.3 Statechart Internals

Statecharts are organized in groups (see Figure 5.1). Following the composite pattern, a statechart is a state itself and can be used as a substate in another state. Transitions connect substates and define the control flow. Every state can be nested into another state to construct state hierarchies. States are equipped with three parameter dictionaries. The first two define the input and output parameters of a state. The third dictionary is used to store local values that can be passed to a substate. States and substates can be compared to classes and instances in object-oriented programming. When the states are defined they are not yet used anywhere, like classes. After instantiation, states are always substates of another state like class instances are usually members of other classes. Transitions between those substates are triggered by events. Transitions do not only specify control flow, but also data flow by attaching a parameter mapping to each transition. This mapping contains instructions on how to fill the input parameters of the next state. Distribution of statecharts over multiple processes is possible by usage of *Remote States*, which transparently represent states located in another process.

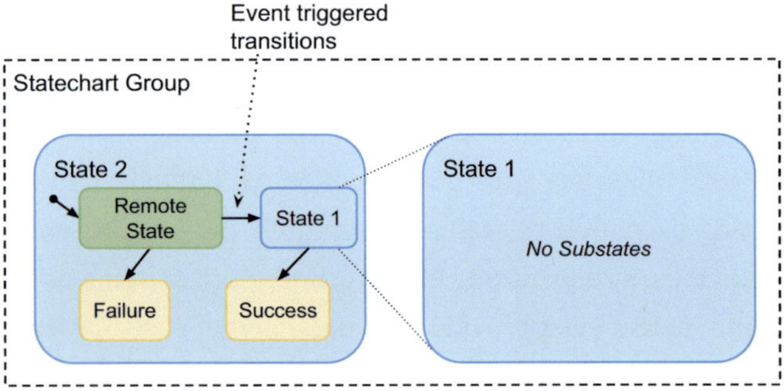

Figure 5.1: Statecharts are organized in groups. States are comprised of transitions and substates. If the statechart groups of a parent state and substates differ, the substate is called a Remote State (green state). The control flow within a state is terminated if any end state (yellow state) is reached. *Source:* (Wächter et al., 2016) © 2016 Frontiers

In the following, the main technical aspects of the statecharts are described: substates, transitions, events, state phases, data flow, interfacing with external components, distributed statecharts and the dynamic statechart structure.
The terms statechart and state are interchangeable since every statechart can also be just a state itself. The term statechart is used if the aspect of representing a state machine is important. The term state is used if properties of states are in the focus.

5.3.1 Substate Types

substates are not the same as states. States are templates which are instantiated as substates of other states. However, only substates of one type are direct instantiations of states. Statecharts consist of four different types of substates, each with a specific purpose:

- *LocalState*
  Local states are normal state instances with no special features.
- *EndState*
  EndStates trigger leaving the parent state immediately. They cannot contain substates or execute any user code. *EndStates* are one way to specify outgoing transitions of the parent state. The name of an *EndState* specifies the name of the outgoing transition of the parent state.
- *RemoteState*
  RemoteStates behave like local states, but internally point to a specific state in another statechart group (potentially in another process or on another host).
- *DynamicRemoteState*
  DynamicRemoteStates are similar to remote states, but they are like generic pointers. On entering, a dynamic remote state morphs into a

specific remote state based on parameters mapped during the transition.

5.3.2 Statechart Groups

Statecharts are organized in statechart groups according to their requirements to external components and by semantic context since all states in one statechart group share the same dependencies. Dependencies are service components that are used by a statechart, e.g. a component to calculate the inverse kinematics of a robot. This means that statechart groups and in turn the contained states will only be started if all dependencies are available. Therefore, the designer needs to be aware of what dependencies are required for a robot skill. Statecharts that use states from other groups, i.e. a *RemoteState*, only share these dependencies implicitly and have only an on-demand dependency on the other statechart group. This means statecharts from a group will not be available only if they are waiting for unstarted remote states. All other states are ready to be executed. This is in particular important for high-level statechart groups such as a statechart group representing planning operators. It is undesirable that the whole system is waiting if one sensor is not available that is only needed for some of the robot's abilities.
Furthermore, the statechart designer should group statecharts in one group by semantic context, e.g. all states needed for directly controlling joints should be in one statechart group.
In general, statechart groups offer all contained states as a service and wait to be used by other statecharts or to be started as a top-level state.

5.3.3 Transitions

Control flow as well as data flow is defined via transitions between states. A transition consists of a source state, a destination state, the associated event and a data mapping that defines the data flow between states during this transition. Each transition is associated with one event that the corresponding source state can process.

Each state has exactly one initial transition if the parent state has at least one substate. The initial transition can be seen as the transition from the parent state to the first substate. This transition is triggered immediately when the parent state is entered. Thus, when the top-level state of a state hierarchy is entered, initial substates on each level are entered recursively until the lowest level of the statechart is reached.

When the control flow reaches an end state the control flow within the parent state is terminated and the associated transition of the parent state is triggered. Thus, each end state defines one outgoing transition in the corresponding parent state.

The data flow during transitions is realized through a parameter mapping definition which is attached to transitions (see subsection 5.3.6). Unlike in Harel statecharts, transitions can only be created between substates of the same parent state to keep the modularity principle of the statecharts. If states had transitions to other hierarchy levels or other parent states, the parent state could not be reused without disconnecting that transition.

5.3.4 State Phases

A state passes the following state phases while it is visited: *OnEnter*, *running*, *onBreak* and *onExit*. Each phase is linked to a user code function, i.e. C++ code, in order to enable developers to execute custom code in a state. *OnEnter*, *onBreak* and *onExit* are atomic coordination phases while *running* is the computation phase of a state, in which complex or long running computations are executed.

The execution order of the phases is as follows: *onEnter*, *running* and then *onBreak* or *onExit*. Before entering a state, i.e. before the phase *onEnter*, the parameters (explained in subsection 5.3.6) are mapped or set to default values. In the *onEnter* phase local variables can be set to be mapped into substates. When a transition is triggered, the *onExit* or *onBreak* phase is entered. Figure 5.2 illustrates when the state phases are executed. Which phase is executed depends on the level where the transition was triggered. Due to the hierarchical nature of statecharts, it is possible for a higher state to receive an event, although its substates have not reached an end state yet. In this case, the substatecharts cannot finish in an expected manner. To give the developer an option to deal with this unexpected behavior, each state provides the *onBreak* phase. If no behavior is specified for the *onBreak* phase, the user code function of the *onExit* phase is executed. When a top-level state receives an event, the complete stack of child-states needs to exit first by exiting all substates, starting with the leaf-substate and proceeding up level by level.

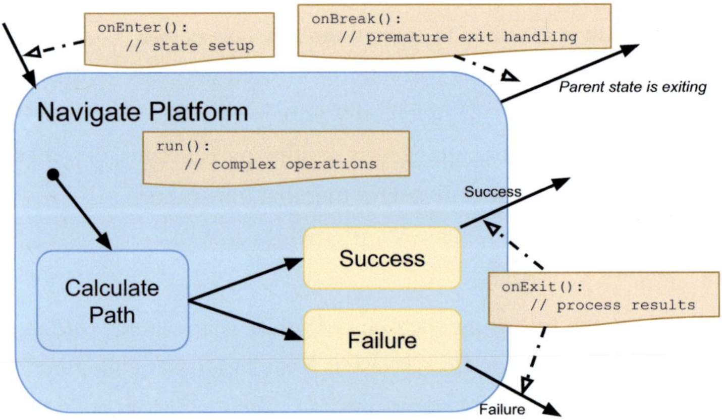

Figure 5.2: Each state visit consists of several phases: *onEnter()*, *run()* and *onExit()/onBreak()*. Each phase is linked to user code.

Whenever a state is entered, its initial substate is entered as well. This means that after executing the *onEnter* phase of a state, the *onEnter* phase of the initial substate is executed immediately afterwards.

Since the user has freedom of implementation in the coordination phases, she or he is discouraged by warnings if computationally intensive code is detected. After entering, the *running* phase is launched in its own thread to allow executing computationally intensive user code without interfering with the statechart coordination. In the default behavior the coordination does not wait for the *run*-function to finish and ignores all results produced by the *running* phase after the state was left.

5.3.5 Events

Transitions can only be triggered by events. Events can be fired either by user code, if an end state is reached or if a certain condition is met. Events from user code or from end states are fired immediately, while events from conditions are fired as soon as the condition is fulfilled. Conditions are specified using Boolean algebra expressions and comprised of literals and Boolean operators.

Event Generation with Conditions

A literal is defined by a data field of an observer and by a parametrized check that is to be performed on this data field. Conditions are installed in sensor-observers and are evaluated by the appropriate observer after each sensor update. To clarify the concept of distributed conditions, the following listing gives an example that will be explained in detail below.

```
Literal objectDistance
    ("ObjectMemoryObserver.hand.pose",
    checks::poseDistance, {object2PoseRef, 10});
Literal forceMagnitude
    ("ForceTorqueObserver.forces.TCP R",
    checks::magnitudeLarger, {5.0});
installCondition ("ObjectReachedEvent",
    objectDistance || forceMagnitude);
```

Listing 5.1: An exemplary definition of an event condition, which fires when either a specific force or a minimum distance is reached.

The first statement in Listing 5.1 defines the literal `objectDistance` that describes the distance between `hand` and `object2` and checks if this distance is below `10 mm`. `object2PoseRef` is a reference to the current pose of `object2` and is updated continuously.

`ObjectMemoryObserver.hand.pose` is the name of the data field for the current pose of the hand within the `ObjectMemoryObserver`. The `poseDistance` check compares the position components of both poses and evaluates to true if the distance falls below the provided argument value (here `10 mm`).

The second statement defines `forceMagnitude` which checks if the force in the right Tool Center Point (TCP) is larger than the given threshold. Both literals are combined using a disjunction. If either of both conditions is true, the corresponding event `ObjectReachedEvent` is fired. The condition is evaluated in a distributed fashion since different sensor observers are used, i.e. `ObjectMemoryObserver` and `ForceTorqueObserver`. A central component called `ConditionHandler` distributes the literals to the appropriate observers. The observer approach avoids unnecessary transmission of high frequency sensor values since only changes of the Boolean state of a literal are signaled by the observers. When the Boolean term of a condition evaluates to `TRUE`, the `ConditionHandler` fires the event associated with the previously described condition. The middleware passes the event to the state that originally installed the condition, which in turn triggers a state transition.

Event processing

Arriving events are queued and processed sequentially by the receiving process. Due to the distributed and asynchronous nature of the software framework, processing of events needs to be performed with caution in order to ensure stability and consistency. One aspect that must be considered, is the fact that a state may already be left when an event arrives. To address this issue, all events contain a unique id of the destination state.

Additionally, special care needs to be taken of consistently considering parallelism. Since statecharts can be distributed over several processes, events

can arrive and be processed in parallel. In order to deal with this situation, the statechart framework protects critical sections allowing concurrent multi-threaded access. Such critical sections are the event-processing function (one per statechart hierarchy-level) and the state phases, where the state coordination is performed (see subsection 5.3.4 for details). Thus, transitions can only be taken once and states are only entered and exited once.
If two events arrive in parallel in two processes (i.e., on two different levels of the statechart), both events are started to be processed. Even though they are processed in parallel a critical section can only be entered by one processing thread. In this example, the lower-level state enters the critical section of its state level[1] first. The upper level, however, traverses down the hierarchy of *active* substates, i.e. states that are currently visited, and encounters that one substate is currently being processed. Since an event is processed in a per-state-level critical section, the upper level state has to wait until the lower level event is processed. Afterwards the upper level event breaks the new active substate and continues upwards, breaking active substates until reaching its own level and exits the currently active substate of its own. The difference between exiting and breaking states is described in the next section.
If the case is encountered where the upper-level state starts processing first and currently breaks a substate and this substate receives an event in the meanwhile, this event is omitted since the receiving state is not active anymore when it acquires the mutex lock.

5.3.6 Transition based Data Flow

One important feature of the proposed statechart concept, which is unique to the author's best knowledge in this extend, is the extensive control of data flow in the statecharts, which eases accomplishing the modularity

[1] State level refers to the hierarchy level of a statechart. Substates of a state are one level lower in the hierarchy.

and reusability principles. All states are equipped with input and output-parameter dictionaries to decouple states from external global data storage. Input parameters are read-only in user code functions and specify all parametrization the state needs for its computations. Output-parameters can be set in the user code functions and contain the results of a state. They can be used as a source for input parameters of the next state or mapped back to the parent's local- or output-parameters.

Additionally, so called local-parameters are provided, which are accessible for the user code. Local-parameters are intended to be used for temporal local storage of parameters that are passed down to a substate's input, passed up from a substate's output or passed between different state phases. During each of the state phases, the developer can access the different parameter dictionaries in the user code functions. Once a state is left, all parameters are reset in order to avoid side effects of previous visits.

Each parameter dictionary field consists of a string identifier and a variant data-type that can contain arbitrary types. ArmarX already provides the basic types Boolean, integer, float, double and string as well as several types associated with robotics, like vectors, matrices, 3D poses or probability distributions. If needed, developers can easily implement new variant types.

These parameter dictionaries are defined by the developer and specify the interface of each state, i.e. which data is needed for execution. Each parameter can be optional, can have a default value[2] and/or can be filled from several sources. This is called *parameter mapping*. When a state is used, its non-optional input parameters without default values need to be connected with other parameters of the same type. Thus, a parameter mapping for each of these input parameters must be created for each state instance. The developer can choose between mappings from the output of a previous state from the same hierarchy level, from the input or local parameters of the parent state or

[2] Consequently, if parameters have a default value, the optional flag does not make much sense any more. Thus these two Boolean flags basically form a tri-state.

from a parameter attached to the transition-event. Additionally, developers can map values from the output of a state to the local- or output-parameters of the parent state. Later, when another substate needs a calculated value as an input parameter, the local parameter is mapped to that input parameter. For example, generic counter states can be implemented following this pattern, so that counting loop sequences of states can be defined without writing any additional specialized custom code. With this pattern, it is possible to pass data from a substate to another state later in the chain more easily. Otherwise the parameters would need to be mapped between all states in the sequence. Figure 5.3 shows the different types of mappings during transitions.

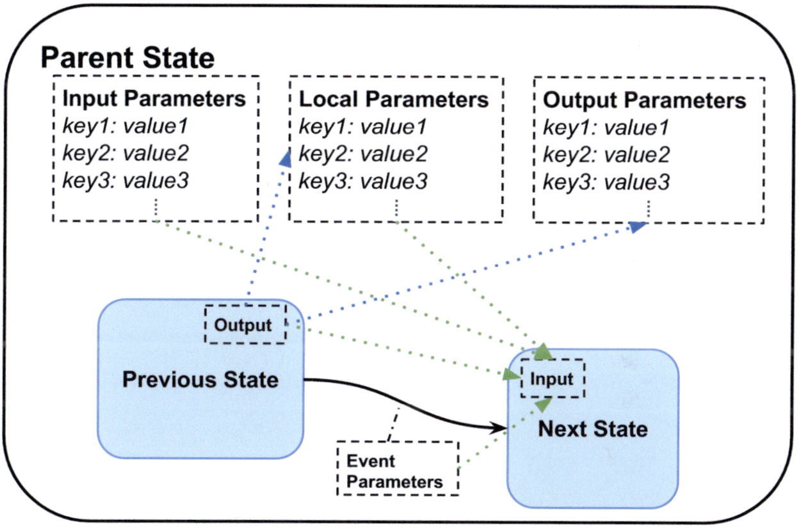

Figure 5.3: Several different parameter mapping sources are possible to fill the input parameters of the next state. *Source:* (Wächter et al., 2016) © 2016 Frontiers

Each state has three parameter dictionaries: input, local and output parameters. In each of the blue substates only the relevant dictionaries for the mapping during the transition are shown. The green arrows show possible mappings to the input parameters of the next state. The blue arrows show possible mappings from the output of the previous state to the local and output parameters of the parent state. These mappings are performed after leaving the previous state and before entering the next state.

Transition Functions

It is often necessary to control the data flow between two states in a more advanced way than by mapping the parameters, especially because states are reusable and might be designed for a different use case. One typical example is that a output parameter of a state is a 6D pose and the input parameter of a following state is a 3D position. In this case, a conversion between these two types is needed. This can be solved with an additional state which realizes the needed conversion. Yet this is cumbersome and increases the complexity of a statechart, since many of these small conversion states might be required. To this end, it is possible to attach a user code function to a transition, which is called directly after the parameter mapping is applied. It gives the developer access to the output parameters of the source state of the transition and to the input and local parameters of the parent state in order to manipulate the input parameters of the destination state of the transition. This way, the range of possible data mappings is greatly extended. Examples of possible data mappings are type conversion, conditional mappings and parameter modification.

This feature significantly reduces the visual complexity of statecharts by removing helper states, which do not contribute to the statechart behavior and are only means to an end. Figure 5.4 illustrates the benefit of *Transition Functions* with an example of parameter conversion needed for the connection of two states. The problem in this example is once solved with an additional state in between and once with a wrapper state. Using wrapper states makes sense if the wrapped state is used multiple times. Without transition functions (Figure 5.4(a)) two additional states are needed to use the `Platform Control` and the `Cartesian Control` state to provide the input parameters in the correct format. The same behavior solved in a more comprehensible way is shown in Figure 5.4(b): Only major states remain and the statechart is not cluttered with helper states. The input parameter conversion has been moved into *Transition Functions*, which are executed during the transition processing.

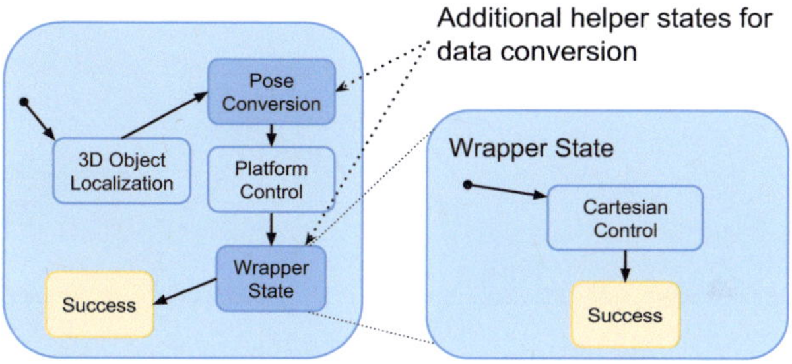

(a) Usage of additional wrapper states

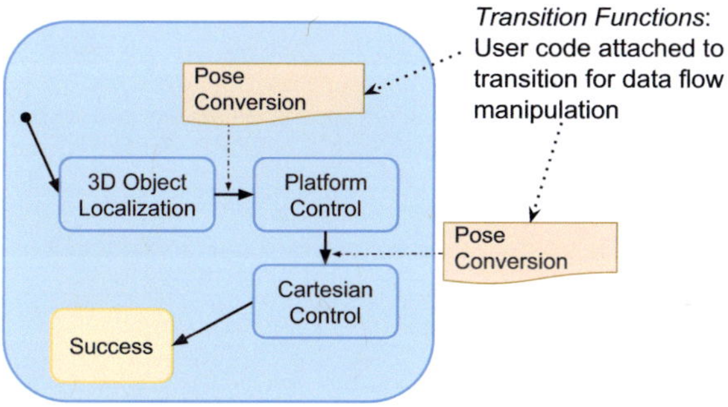

(b) Usage of transitions functions

Figure 5.4: *Transition Functions* reduce the number of needed states and improve the comprehensibility of reused statecharts.

5.3.7 Statechart Profiles

Another extension of the Harel's statechart concept, called *Statechart profiles*, offers the possibility to provide different parameter sets for different robots or different setups. These profiles are defined in a hierarchy tree, which means that values of lower priority profiles (lowest priority is the *Root*-profile) are overwritten by values of a profile with a higher priority. This hierarchy allows using inheritance of parameter values to avoid duplicated parameter values. Figure 5.5 shows an example of such a hierarchy for the ARMAR-III robot (Asfour et al., 2006).

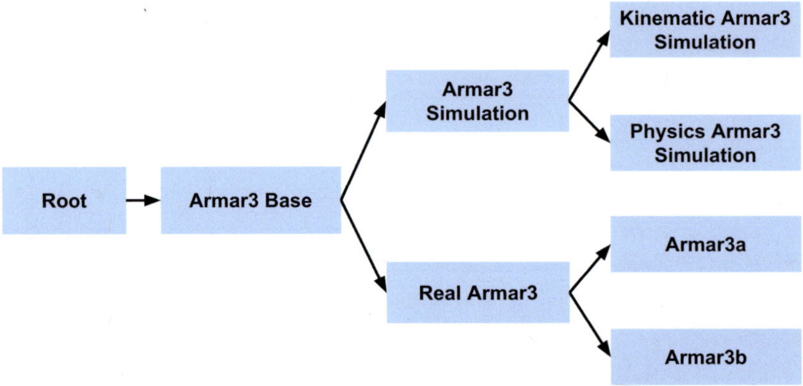

Figure 5.5: Statechart profiles allow for individual parameterization of each state for a specific (robot) use case and are organized in a hierarchy tree. To avoid repeated specification of shared parameters for each use case such parameters can be specified in the fitting parent profile of the profile tree.

Default State Parameter Values

A profile encapsulates default parameter values for a well-defined use case configuration of a statechart while maintaining the same functionality for all statecharts of the hierarchy. For example, a statechart that is used on a simulated robot might need different default parameter values than the same statechart that is deployed on a real robot. Thus, the usage of statechart profiles helps preventing duplicates of functionally equal statecharts while still allowing convenient configurations for different use cases.

Statechart Group Default Configuration

Additionally, statechart profiles allow specifying configuration values for the whole statechart group like proxy names, e.g. the name of the KinematicUnit to control the robot's actors. These names usually differ from robot to robot and need to be specified for each robot separately, e.g. *Armar3KinematicUnit* for the robot ARMAR-III.

5.3.8 Interfacing with External Components

Statecharts that can only access functionality and data of themselves are not particularly useful for robotics. Therefore, they must be able to access all available components. Since ArmarX is a distributed system, it cannot be assumed that required components are running in the same process or on the same host. Hence, states require network proxies to these components. Additionally, it should be ensured that a state is only started if all required components are available. Dependencies for a group of states can therefore be defined in a so called *StatechartContext*, which manages dependencies and enables states to communicate with external components.

5.3.9 Distributed Statecharts

Another important feature is the possibility to distribute statecharts over several processes or hosts. To this end, states are organized in groups, which for example contain states that are semantically similar and share the same dependencies to external components. In this context, semantically similar means states which share common aspects regarding their purpose. For example, all states for controlling holonomic platform movements from a PD controller to path navigation should be encapsulated in one statechart group. However, this is just a useful convention.

Each group is executed as one component in a so called *RemoteStateOfferer*. These *RemoteStateOfferers* offer states to be used by other states as *RemoteStates* over the network. *RemoteStates* are network proxy objects, which acts like a local pointer object. For increased robustness, each *RemoteStateOfferer* is located in its own process. Thus, a *RemoteState* is employed whenever a state uses a state of another group as a substate. This process is completely transparent to the developer. The only difference to a local state is that the *RemoteStateOfferer's* name needs to be specified in addition to the state name. Theoretically, each state could have its own group for maximized robustness. Since distributed statecharts are slower than local statecharts, developers need to decide carefully when to split statecharts into more than one group.

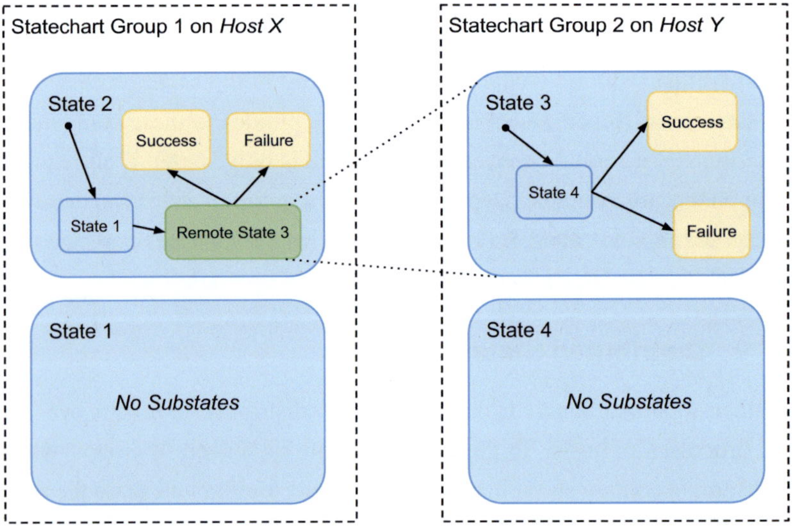

Figure 5.6: Statecharts are organized in groups which can be distributed over several processes and hosts. By default, each statechart group resides in one process. It is possible to incorporate states of another group transparently into a statechart by using *RemoteState* instances. *Source:* (Wächter et al., 2016) © 2016 Frontiers

Another advantage of distributed statecharts is the possibility to deploy them close to their components. A statechart that makes heavy use of the robot's memory should ideally be located on the same host as the database servers,

whereas a visual servoing statechart should be close to the vision system and to the host where joint-level control takes place. Figure 5.6 depicts the linkage between different statechart groups and *RemoteStates*.
Due to the sophisticated underlying middleware Ice[3], which transforms network communication into transparent function calls, the step from local statecharts to distributed statecharts only requires to provide interfaces in the IDL of the middleware and an implementation of these interfaces. Substates pointing to a remote state just use another implementation of the state interface, which reroutes all the function calls over the middleware. On the other side, the aforementioned *RemoteStateOfferer* component offers a network interface to the normally local functions of a state. This way, consistency is assured in the same way as it is done locally, with mutexed access and holding of data only on the offerer side. Thus, synchronization of data is not needed.

5.3.10 Dynamic Statechart Structure

In most statechart frameworks, the structure of the statecharts is fixed, once it has been designed by the developer. This limits the usability of statecharts in a highly dynamic environment, e.g. regarding humanoid service robots. In this context, a symbolic planner may be incorporated which needs to be able to change the statechart structure on the fly, according to the currently planned program flow. ArmarX supports dynamic online statechart restructuring by offering so called *DynamicRemoteStates* which provide generic entry points for exchangeable statecharts. As the name suggests, a *DynamicRemoteState* connects to a state in another (or its own) process. It decides upon entering, into which state it is morphed based on specific parameters passed by the transition. Additionally, further parameters can be specified that are mapped into the connected state. The correctness and completeness of the parameters are verified at runtime, i.e. when the state is loaded.

[3] `https://www.zeroc.com`

5.4 Formalization of the Statechart Concept

The full statechart concept is formalized as follows. All capital letters, if not specified otherwise, are sets of elements named with the same letter in lowercase, e.g. $S = \{s_1, \ldots, s_k\}$ is a set of states s.
A statechart group is defined as the tuple

$$G = (S, O, C) \tag{5.1}$$

which consists of a set of states S, a set of proxy objects O, which the states can use to communicate with external components, and a set of configuration key-value pairs

$$C = \{(key_1, value_1), \ldots, (key_n, value_n)\}. \tag{5.2}$$

The tuple

$$s = (id, I, T, i_{start} \in I, E, P, F) \tag{5.3}$$

defines a state s, where id is a unique identifier, I is a set of state instances and T a set of transitions. i_{start} is the initial state instance and an element of I. E is a set of transition defining events, P is the tuple (P_i, P_l, P_o) and F the tuple

$$\begin{aligned} F =&(f_{enter}(P_i, P_l, P_o), \\ &f_{run}(P_i, P_l, P_o), \\ &f_{break}(P_i, P_l, P_o), \\ &f_{exit}(P_i, P_l, P_o)), \end{aligned} \tag{5.4}$$

where $f_{enter}(), f_{run}(), f_{break}(), f_{exit}()$ are user code functions associated with state s. P_i, P_l, P_o are the input, local, and output parameter dictionaries.

A state instance i is the tuple

$$i = (s, n), \tag{5.5}$$

where $s \in I$ is a state, the central component of the statechart concept, and n the identifier of the state instance. State instances always belong to a set I of state instances of another state.
A transition $t \in T$ is defined by the tuple

$$t = (M, e, i_s, i_d), \tag{5.6}$$

in which M is a set of parameter mappings m, e is the event associated with the transition and i_s and i_d are the source respectively destination state instances of the transition. A parameter p is the tuple

$$p = (key, type, optional, default), \tag{5.7}$$

where key is a string identifier, $type$ is the type of this parameter (potential types: Boolean, int, string, float, double, position, orientation, pose, ...), $optional$ is a Boolean flag to signal whether or not this parameter is required for successful execution and $default$ is the default value of the parameter with the type of $type$. Default values are stored in the implementation of this statechart concept in JavaScript Object Notation (JSON) syntax[4].
The parameter mapping m is the tuple

$$m = (mappingtype, sourcekey, destinationkey), \tag{5.8}$$

in which $mappingtype \in \{output, parent, value, event\}$ determines from which source a parameter is retrieved with the $sourcekey$ and inserted in the input parameters of the destination state at key $destinationkey$.

[4] `http://json.org`

The statecharts in the ArmarX framework are implemented based on this formalization.

5.5 Textual Statechart Specification

While the advised method to create statecharts is to use the graphical statechart editor (see section 5.6), it is also possible to specify statecharts textually as shown in Listing 5.2. First, each state needs to be added with its state class (TemplateParameter) and the instance name (parameter of *addState()*) as a substate (`addState()`) to link the substate with the user code of the added state. Afterwards, transitions between these substates can be created by specifying the start and end state and by declaring on which event this transition should be triggered. The data flow and any other feature can also be specified, but these specifications are omitted here for brevity.

```
void defineSubstates()
{
  setInitState(addState<InitialState>("Initial"));

  StateBasePtr finalSuccess =
      addState<SuccessState>("Success");
  StateBasePtr finalFailure =
      addState<FailureState>("Failure");

  addTransition<Next>(getInitState(),
      getInitState());
  addTransition<TimerExpired>(getInitState(),
      finalFailure);
  addTransition<Success>(getInitState(),
      finalSuccess);
}
```

Listing 5.2: An exemplary textual definition of a state.

5.6 Graphical Modeling of Robot Skills

Since statecharts are a visual modeling approach, the next logical step is to program statecharts with a graphical user interface. This eases the comprehensibility of statechart designs immensely and increases the development speed drastically since repetitive textual specifications are avoided.
For the framework ArmarX, a powerful graphical statechart editor has been developed based on the concept presented in this chapter, which allows the developer to design hierarchical statecharts by specifying the control and data flow of robot skills in an intuitive way. The goal of the statechart editor is to enable users with little training to create event-driven robot behaviors or skills from scratch or by reusing existing skills such as moving a platform or complex skills such as grasping an object. When existing skills are reused, it is possible to create more complex skills without programming any line of code by just coordinating data flow between existing skills.

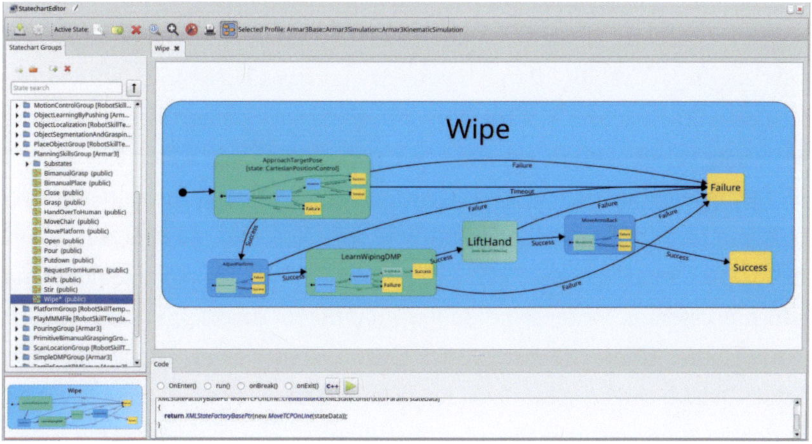

Figure 5.7: With the graphical statechart editor it is possible to design, to manage and to execute robot skills.

The statechart editor provides all functionality for designing statecharts, from creation and organization over hierarchical arrangement to linking the states with user code. Figure 5.7 shows a screenshot of the graphical statechart editor, which is explained in detail in (Wächter et al., 2016). On the left, the organizational structure of statechart groups is shown as a tree. This structure is important for the execution of statecharts since one statechart group is always executed in one process. The main part is the view of the currently opened statechart including the completely visible hierarchy of the statechart. The bottom shows the user code linked with the currently selected state.

5.7 Summary

In this chapter, a new robot skill modeling approach was presented that extends the original statechart formalism proposed by (Harel, 1987) to the requirements of robotics. The presented statechart concept allows for hierarchical modeling of skills on all abstraction levels while supporting and encouraging reusability through defined interfaces between states and precise data flow control. Furthermore, a defined success and failure result of each state ensures failure handling on coordination level, which can be done on the appropriate parent level. Distributing statecharts transparently over multiple hosts addresses the multi host solutions that are typically found in complex robot systems such as humanoid robots. Another requirement, which is essential for reusable skills with different robots, is solved by introducing hierarchical statechart profiles that allow specifying default parameter values, which can be specialized for specific robots. The complete approach is tightly integrated into the robot development environment ArmarX and represents a core feature of the framework.

6 Task Solving and Execution in Dynamic Environments

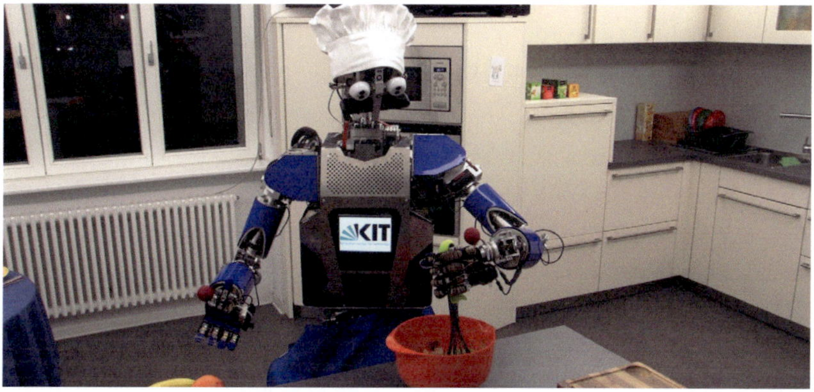

The developed methods for understanding and representation of human demonstration allow for execution of observed tasks as a sequence of actions, which are specified by statecharts with predefined parameters. But tasks in human-centered environment are highly dependent on the current state of the world. Depending on this state a different sequence of actions might be needed to solve a task. For this reason, it is not sufficient for an autonomous humanoid robot to only be equipped with a fixed set of observations, which is a specific action sequence to solve one task in one configuration. To this end, an approach is proposed to solve complex tasks in a perceived, dynamic environment. This approach includes fault recovery and affordance reasoning based on a given goal and various knowledge bases about the environment. Parts of this approach were published in (Wächter et al., 2018).

6.1 Task Execution Overview

The proposed task execution approach contains three main modules: The domain and plan generation module, the symbol replacement manager and the task execution and monitoring module. The approach is depicted in Figure 6.1 with its main modules and their control flow and data flow connections. To motivate the approach consider the following example: The human utters the following request to the robot *I'd like to drink some soda. Could you bring me some soda?* This spoken command is processed by a Language Understanding component (Ovchinnikova et al., 2015) and translated into a symbolic goal with affordances for each object uttered in the command (Wächter et al., 2018), i.e. `inHand(humanHand, soda)` as the goal and the affordance *drink* for the *soda*. Affordances (Gibson, 1979) describe the action and interaction possibilities of an agent on the environment. This data is transferred to the Symbol Replacement Manager, where a preliminary feasibility check on the involved objects is performed, i.e. it is examined whether all objects are known and an object instance is already located in the current world state for each object. In this example, it is checked whether or not *soda* is a known object and if a location hypothesis[1] can be generated for it. *Soda* is unknown to the robot as a graspable object[2] and, thus, the robot looks for a suitable replacement with its available *replacement strategies*. Replacement strategies employ different kinds of knowledge bases or sensor input for object replacement candidates or location hypotheses, e.g. common places of objects experienced by the robot in the past (Kozlov, 2013).

[1] *Location hypotheses* are guesses of the robot where an object in the real world might be. These guesses are needed since classical planners need complete knowledge about the world.

[2] In this work, *graspable objects* are objects of which the robot knows how to grasp them and whom the robot is able to visually localize.

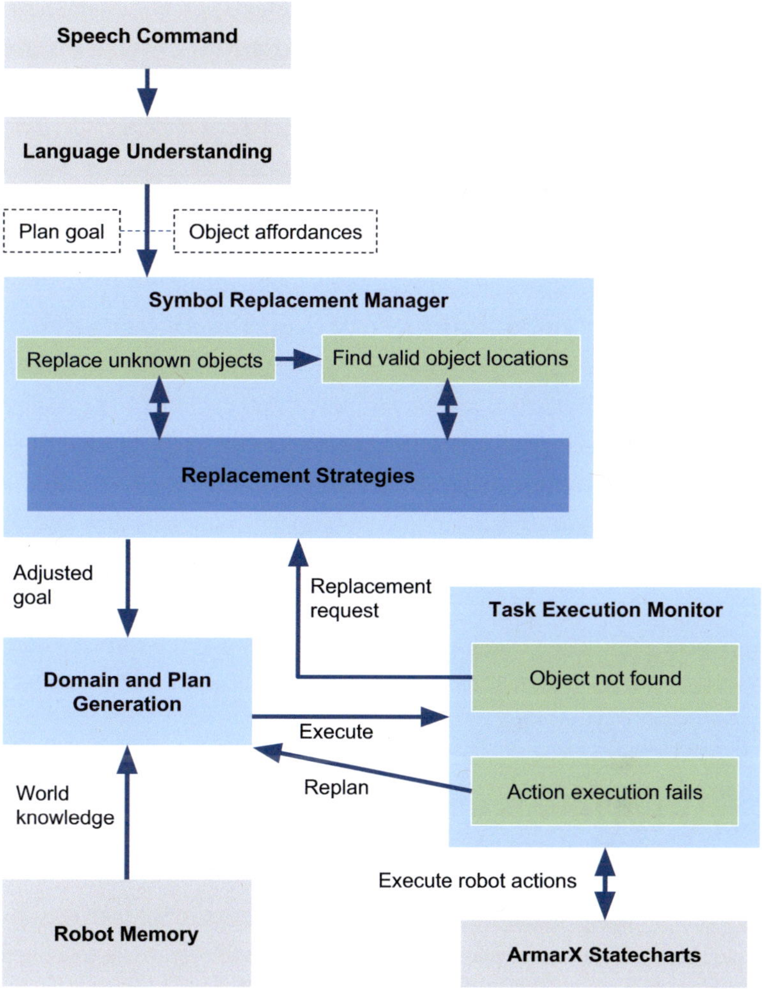

Figure 6.1: The main modules, control flow and data flow of the task execution approach. The green and blue components were developed in this work.

Since *soda* is unknown to the robot as a graspable object, it does not need to find a location for it. Yet the robot can try to find another object that might fit the task. Based on the affordance *drink*, which was provided by the Language Understanding component, it searches through the common sense database for an object that has the affordance *drink* associated with it. It finds the object *juice*, which is a graspable object for the robot. If all objects have a location hypothesis or are replaced with objects that have one, the adjusted goal is passed to the Domain and Plan Generation module.
The Domain and Plan Generation module transforms the continuous robot experience into symbolic predicates and generates a domain and problem statement for the planner. Based on this, the planner tries to find a sequence of planning operators that solves the given problem, i.e. achieves the goal state. If a solution in form of a plan has been found, the sequence is executed sequentially by using a link between symbolic planning operators and state-charts, which execute the planning operators on the robot. The execution is monitored with respect to perceived and expected effects of each action, e.g. the juice might not have been grasped successfully and it is still standing in the fridge. If a mismatch is detected, a new plan is generated based on the current world state and the execution starts over.

6.2 Continuous and Consistent Robot Knowledge

The first step towards online planning on a humanoid robot is to transform the perceived, continuous world state into a symbolic representation suitable for task planning. The raw sensor data as well as subsymbolic information such as object positions need to be translated into symbolic predicates like *the apple is on the table*. However, this is a complex endeavor due to the ambiguity, uncertainty and lack of data. The uncertainty about objects and their location requires a higher tolerance for predicates like *the apple is on the table*, which in turn leads to ambiguity of the predicates, because they might

overlap. These issues make the transformation of the robot's knowledge into a symbolic world description difficult and require mechanisms to reduce uncertainties and assumptions. Another issue is the impermanence and general uncertainty of perceptions: Every state of the art object recognition sometimes provides false positives and inaccurate localization. Additionally, an object that was perceived some time ago might have been moved by a third party. All these issues need to be considered in a robot memory architecture, which represent what the robot "knows" about the world.
With these requirements in mind, the robot memory framework *MemoryX* has been designed and realized, which builds the base for the approach presented in this chapter. The details of *MemoryX* are explained in subsection 2.4.1. In the next section, the approach of using the robot's memory for reasoning about object and location replacements is presented. In section 6.4, the use of the robot memory structure to extract the needed symbols for the planner is described.

6.3 Symbol Replacement based on Various Knowledge Bases

Reasoning about entities in the environment of the robot is an important ability for an autonomous robot to solve new and unknown tasks. In this thesis, a symbol replacement[3] system is presented which has the ability to reason about the suitability of objects for a given task and proposes replacements based on various knowledge bases and sensor information. Further, the approach can generate location hypotheses for objects that have not been seen recently or could not be found anymore at the last known location. The location hypothesis generation is a requirement for the usage of classical planning systems in dynamic and unexplored environments. Classical planning systems (see section 2.5) need complete world knowledge to solve a

[3] Symbol replacement is here the replacement of planning symbols like objects or locations in the planning domain.

task. They cannot reason on unknown and incomplete world states, e.g. all objects involved in the task and their locations need to be known in advance. To avoid actively searching with the robot for all objects before starting to plan, location hypotheses are generated for all objects. This means, object instances are inserted into the working memory at assumed locations with a relatively low existence certainty and high position uncertainty. Only during execution these hypotheses might be proven wrong and will trigger a replanning with the new information that a particular object is *not* at a specific location. One side product of the symbol replacement is a feasibility check of a given task, i.e. whether all objects mentioned in the goal are known to the robot. Otherwise the planning system will not be able to find a solution.

6.3.1 The Replacement Process

Figure 6.2 shows an overview of the symbol replacement system and Figure 6.1 shows how it is used in the planning system. The Symbol Replacement Manager (SRM) is evoked by an external trigger like the Language Understanding component or if a plan fails during execution. For each object and location name occurring in the goal provided by the external trigger, the SRM checks if they are contained in the robot's memory (*MemoryX*), i.e. if the names can be converted into *MemoryX* types that have instances with specified locations. If an instance or its location is missing, the SRM attempts a replacement and rewrites the goal before passing it to the planner. The SRM queries the domain generator and replaces unknown objects in the goal with known ones and makes sure that location hypotheses are available for instances of all objects mentioned in the goal by inserting object instances into the working memory. The planner will treat these in the same way as confirmed object instances, but all actions using this object instance will fail during execution if the object instance hypothesis is wrong.
To find replacements a strategy pattern is used. Several replacement strategies are available, which are associated with an overall strategy confidence

to determine the order in which the strategies are called. The result of a replacement strategy is a list of replacement propositions with a confidence value for each replacement hypothesis. For example, the common location strategy, i.e. locations extracted from experience, has a higher confidence than the common sense location strategy, which generated object locations from data-mining of large text corpora and potentially contains more false propositions than grounded knowledge from experience. The strategies are explained in subsection 6.3.2.

If a suitable replacement has been found, the SRM rewrites the goal and passes it to the planner that generates a plan. The plan execution is supervised by the Plan Execution & Monitoring component. If the plan execution fails because of a missing object, the SRM is called again.

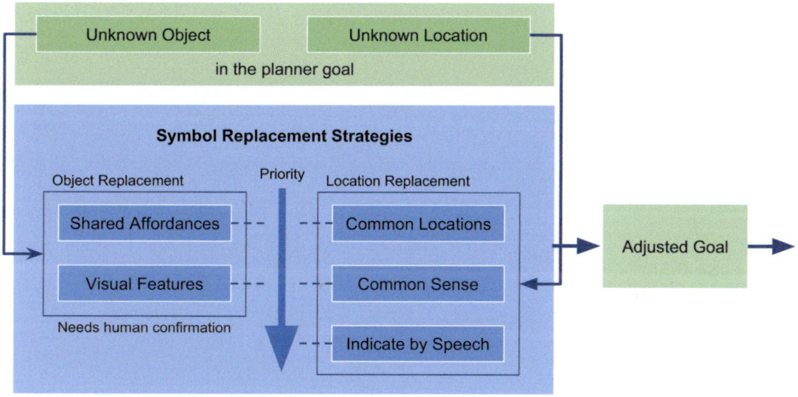

Figure 6.2: The Symbol Replacement Manager provides replacements for objects and locations based on various replacement strategies. These strategies include replacement based on shared visual affordances or on common sense knowledge extracted from large text corpora.

6.3.2 Symbol Replacement Strategies

The Symbol Replacement Manager is using different replacement strategies, which vary with respect to the considered input data and which range from online visual shape estimation to offline evaluation of large text corpora. The replacement strategies are subdivided into object and location replacement strategies, which are described in the following.

Object Replacement Strategies

Object replacement is performed when a) an object type mentioned in the goal is unknown or b) a suitable object could not be found at any known location of the object during the plan execution. Two object replacement strategies based on shared common sense affordances and shared visual features as described below are presented here. Object replacement requires human feedback. Therefore, the SRM generates a confirmation question for the human and only proceeds with the replacement if this question is confirmed.

Common Sense Affordance Strategy

The common sense affordance strategy uses shared affordances between objects extracted from large text corpora as presented in (Kaiser et al., 2014). For each object known to the robot[4] affordances expressed by verbs combined with a confidence score are extracted and stored in a database. Based on patterns like "`VERB (a|the)? NOUN`" (*cut the banana*) or "`VERB with (a|the)? NOUN`" (*cut with a knife*), the relations between objects and affordances are extracted. Based on the occurred pattern, the *role* of the object is determined, i.e. *patient* or *instrument*. The frequency of occurrences determines the confidence score of the relation. This results in the tuple ⟨*object, affordance, role, score*⟩, e.g. ⟨*knife, cut, instrument, 0.823*⟩.
The resulting affordance database is then further processed to generate a replacement database containing tuples of the form ⟨*object1*, *object2*, *affordance*, *score*⟩. This tuple indicates that *object1* can be replaced with *object2* in context of the *affordance* with the confidence *score*. These tuples are gen-

[4] Only known objects are extracted since the robot cannot handle unknown objects and therefore the proposition of unknown objects is not expedient.

erated with a relational learning framework as described in (Wächter et al., 2018).

Since the replacement database is generated from textual corpora, it contains only object names represented by nouns ("cup" instead of "bluecup"). The strategy uses the type hierarchy (e.g. "bluecup" $\rightarrow$ "cup" $\rightarrow$ "container") for realizing the replacement. For example, if "bluecup" needs to be replaced, it will search for replacement options specified for its parent in the hierarchy. Similarly, if "cup" is suggested as a replacement, the strategy will select its leaf child in the hierarchy as a replacement candidate.

Shared Affordance from Visual Features Strategy

This strategy uses shared affordances between objects extracted from visual features to suggest replacement objects. This strategy was developed in (Mustafa et al., 2016). The visual features are calculated based on segmented point clouds of the objects and utilize the shape of the objects. A global 3D histogram descriptor is generated to estimate the affordance of an object (Mustafa et al., 2015). In Figure 6.3 the perceived point cloud is projected into the visualization of the current working memory content of the robot, where for each segmented object (colored point cloud clusters) the estimated affordances are shown. In this example, the bowl as well as the basket (the object on the right) afford *pouring into*, *stirring* and *dropping into*. The grey bowl represents the perceived pose of the bowl by the object recognition system. The perceived object is compared to an affordance database that was learned by the JointSVM algorithm (Mustafa et al., 2015) on labeled point cloud data of objects.

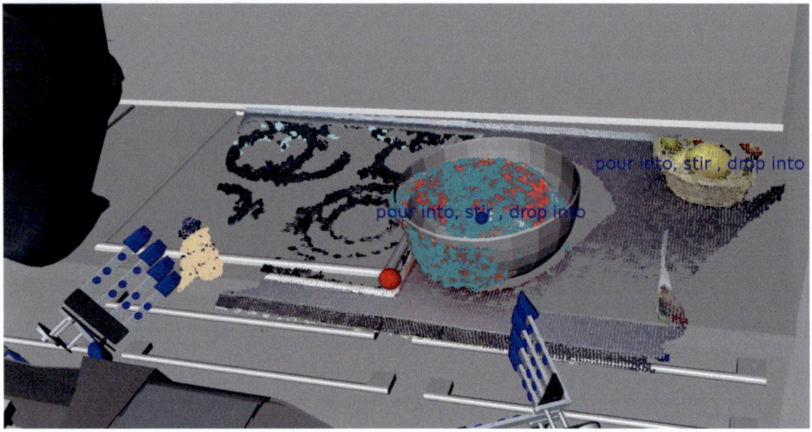

Figure 6.3: Affordances estimated based on shape representations. *Source:* (Wächter et al., 2018) © Elsevier 2018

Location Replacement Strategies

Location replacement happens when the locations of all instances of an object type mentioned in the goal are unknown or when an object could not be found during the plan execution. For location replacement, the SRM manipulates the current working memory of the robot and inserts object instances as unconfirmed hypotheses at the suggested location provided by the replacement strategy. Three location replacement strategies are presented in the following paragraphs, which are based on 1) common locations learned from the previous experience of the robot, 2) common sense locations obtained from textual corpora, 3) human feedback.

Common Locations Strategy

The strategy based on common locations uses a feature of *ArmarX* that allows robots to learn typical locations of the objects from their experience (Welke et al., 2013b). Since the information about object locations is stored in the robot's memory as a set of density distributions of points (see Figure 6.4), the distributions have to be mapped to symbolic location labels that can be processed by the planner. To do so, a location label is linked with the expected value of the corresponding distribution. When the robot is close enough to the object and can actually see it, the assumed location is corrected with the actual observed object position.

This strategy is the most used location strategy with a high hypothesis confidence. The confidence of the location hypotheses is rated high because the knowledge originates from the experience of the robot. When the robot is started, the locations of the objects are unknown. Thus, when a command is received, a location hypothesis needs to be generated for each implied object.

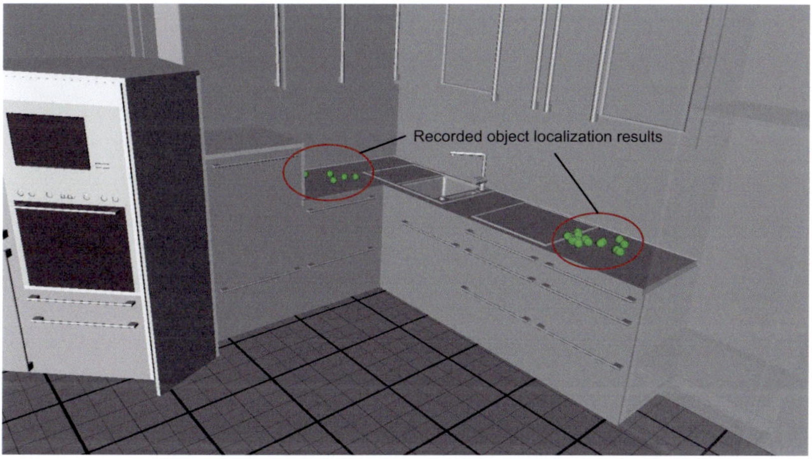

Figure 6.4: Previously seen locations of an object (green spheres), which are clustered to density distributions in order to generate common object locations. The cluster on the right side represents the top location hypothesis, since the object has been seen at this location more often. *Source:* (Wächter et al., 2018) © Elsevier 2018

Common Sense Locations Strategy

The strategy based on common sense knowledge is obtained from textual corpora and employs the method for extracting typical object locations from text as described in (Kaiser et al., 2014). For each pair of (*object_label*, *location_label*) in the underlying domain, such that the labels are expressed by nouns, the corpus is searched for the patterns "`OBJECT_NOUN (be)? loc_prep LOC_NOUN`", where `loc_prep` is a location preposition, e.g. *on*, *in*, *at*. Using this method, a location database is generated consisting of the tuples of the form ⟨*object*, *location*, *score*⟩. Each tuple proposes a location for a given object, with a confidence score corresponding to the normalized frequency of their co-occurrence in the text corpus. To increase the likelihood of finding objects, this strategy queries the database not only for the actual object type, but for all parents of that object type in the type hierarchy.

Human Feedback Strategy

The strategy based on the human feedback uses object locations communicated by the human, e.g. *The corn is in the fridge*. The Natural Language Understanding component handles world state descriptions including location descriptions and updates the working memory in *MemoryX* correspondingly. Thus, the strategy consists of generating a question for the human, asking for the location of the missing object, and monitoring the working memory updates in *MemoryX*. Once *MemoryX* has been updated, the SRM invokes the planner, which replans based on the new information. This strategy is considered a fall-back mechanism that is used only if all other strategies fail.

6.4 Symbol Extraction from Continuous Robot Knowledge

Symbolic planning requires that the world is represented purely by symbols and not by subsymbolic data. However, representing the robot's environment as symbols perceived by the robot itself is a difficult endeavor. Table 6.1 shows an comparison between subsymbolic and symbolic representation of robot knowledge.

Each symbol type may require different (sensor) data and algorithms for its calculation, e.g. the predicate `graspable` queries the database for grasping knowledge while the predicate `on` evaluates the position of object instances in the working memory. Therefore, a strategy pattern is used for generating predicates in a uniform manner. For each predicate a specialized predicate provider is available that can access each memory of *MemoryX* or any other robot component like the current robot model state. Predicate providers are free in their realization and can be simple rules such as `distance(object,hand)` $\rightarrow$ `inHand(object,hand)` or learned from experience. For example, the `graspable` predicate could be refined after each grasp trial with the result of the execution. Additionally, predicate states can be stored in the relation segment of the working memory. This is used by memory updates from natural language, e.g. *the fridge is open*, or for non-observable action effects (see section 6.7).

Table 6.1: Comparison between subsymbolic and symbolic representation of the robot knowledge.

	Subsymbolic	**Symbolic**
Modeling Space	$\mathbb{R}^n$ per type	Types (e.g. `object`) Instances (e.g. `bottle`) Hierarchies (`bottle` is `container`)
Operators	Arithmetic Matrix Operations	Predicates Constrained Rules
Domain Description	Lists of $\mathbb{R}^n$ per type	Instances Predicates Rules

Every time the planning domain is generated, all predicate providers are queried to provide their predicates given the current world state. The validity of the predicates highly depends on the accuracy of the used data. If an object position has an average error of a few centimeters, the threshold regarding objects needs to be chosen more tolerant, which opens the door for false positives. For example, the `inHand` predicate on the ARMAR-III robot compares the distance between object instance and hand to determine whether or not an object is in the robot's hand since it has no sensors in its hand. Yet visual in-hand localization is not robust due to occlusion and the motion prediction of the object is not always correct. These issues require a threshold of several centimeters for the `inHand` predicate, which leads to false positives if the hand just happens to be close to the object. The predicates used in the proposed approach are described in subsection 7.3.1.

6.5 Generation of Symbolic Domains from Robot Knowledge

Figure 6.5 shows how the robot's memory is used to generate a symbolic domain description consisting of static symbol definitions and problem specific definitions. The symbol definitions consist of types, constants, predicate definitions and action descriptions, while the problem definitions consist of the symbolic representation of the current world state represented by predicates and the goal state that should be achieved. Types enumerate available agents, hands, locations and object classes contained in the prior knowledge. Constants represent available instances, on which actions can be performed, and are generated using entities in the working memory. Each constant can have multiple types, such that one is the actual type of the corresponding entity, and others are parents of that particular type including transitive parentship. For example, instances of the type `cup` are also instances of `graspable` and `object`. This type hierarchy is important for specifying actions over a particular set of types, e.g. the grasping action has the type `graspable` as a parameter to ensure that grasps are only planned on graspable objects.
The domain generator derives action representations from the long-term memory, where they are associated with specific robot skills represented by statecharts (Wächter and Asfour, 2015). Each action is associated with a set of preconditions and effects represented by predicates and predicate changes respectively.

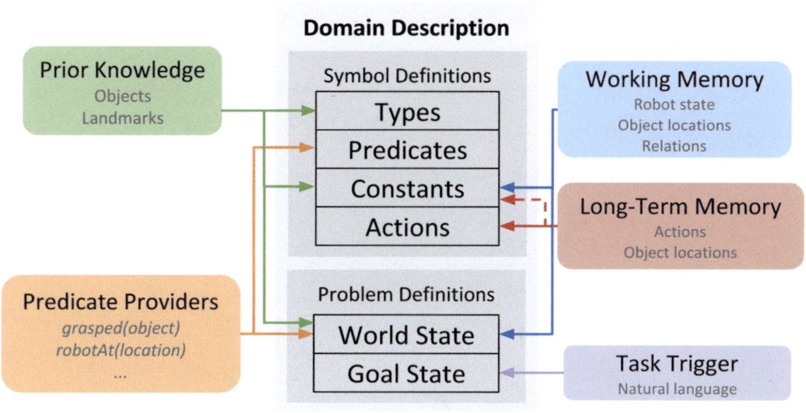

Figure 6.5: Components involved in the domain description generation.

The generated domain description is used by the replacement manager and the planning component. The replacement manager uses the description to check if objects in the planner goal and their locations are known to the robot. The planning component uses it as the knowledge base for finding plans.

6.6 Execution of Symbolic Planning Operators

Symbolic planning operators only contain information needed by the planner, i.e. preconditions and effects. To execute a planning operator on a robot an entirely different representation is needed. This representation needs to represent actions on a symbolic level as well as to perform these actions on a technical level, e.g. move a robot hand along a trajectory. In this thesis, the formalization *Object-Action Complex* (OAC, see section 2.2) is employed to connect the symbolic world with the execution on a robot. The prediction function of an OAC is represented by the preconditions and effects of the planning operator. The execution ID is a link to a statechart, which contains all the details required for execution. The statechart for the planning operator

`PlaceObject` is shown in Figure 6.6. The success measure and learning parameters can be queried and updated by the statechart from the OAC segment in the long-term memory of the robot. The learning parameters stored in the database can be used to improve the performance of an action.

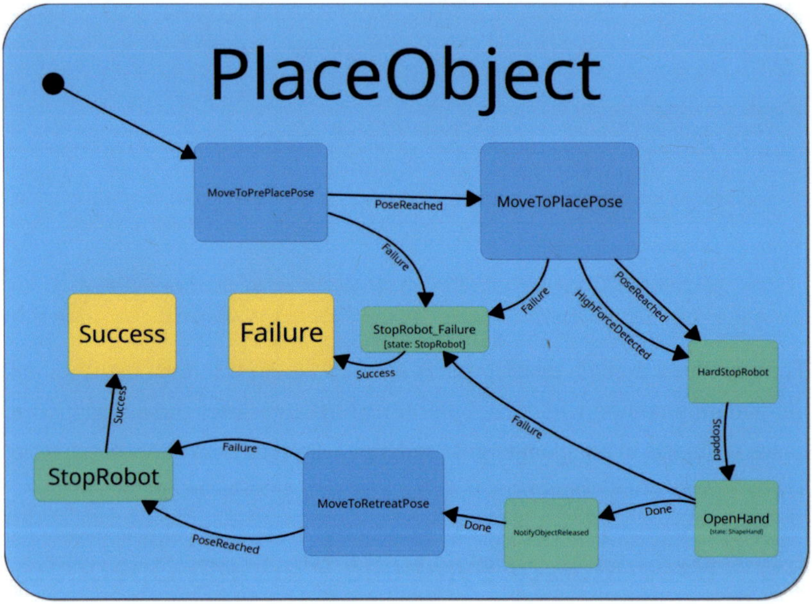

Figure 6.6: The equivalent of the planning operator `PlaceObject` represented as a statechart for execution on a robot. Only the first level of the hierarchical statechart is shown for simplicity.

The statecharts developed in this thesis are designed to fulfill the requirements of a planning operator realization. It is easily possible to create statecharts, which accept symbolic parameters like a planning operator as input. Due to the hierarchical nature of statecharts, the required levels of abstraction can be modeled, from a PD controller for motion control of the mobile platform of the robot to the highest abstraction, the planning operators.
All parameters of a planning operator are entities of the memory since the planning domain has been generated from the robot's memory. The statechart

framework is capable of processing this entity type and further information, which is required for executing an action, can be queried with these entities from the memory. For example, the planner only needs symbolic information of an object instance, whereas statecharts need the exact 6D location and grasps associated with the object.

Similar to the applicability of a planning operator, a statechart can be inapplicable on a specific world state. Yet the reason of inapplicability, which will lead to a failure of the action, can be manifold and can only be determined during execution. One reason for inapplicability of a statechart is that an object could not be found at the location hypothesis. The next section explains how such failures are handled by the system.

6.7 Task Execution and Monitoring

Once a solution, i.e. an action sequence, has been found by the planner it needs to be executed. Furthermore, instead of blindly executing the actions one by one, the execution should be monitored and verified for correctness and success. Figure 6.7 shows the control flow during the execution. The Task Execution Monitor (TEM) receives the solution from the planning component and associates each planning operator with a statechart with OACs from the long-term memory. Before the execution of each action, the applicability of the action is verified again on the level of the planning operator's preconditions. If the action is still applicable on the current world state, the execution of the associated statechart is triggered. After the execution of each action the effects of the planning operator are compared to the changes in the robot's memory perceived by the sensors of the robot. If all observable effects are fulfilled, the task execution is still on track and the next action can be executed. However, not all effects are observable by the robot. For example the state of the fridge door cannot be perceived in the current system. Such action effects are not validated. Yet these effects are tracked in the robot's memory by using the effect descriptions of the planning operators since they might be a requirement for the next task. To account for failures regarding these action effects the human can inform the robot about all states, i.e. effects, using natural language feedback.

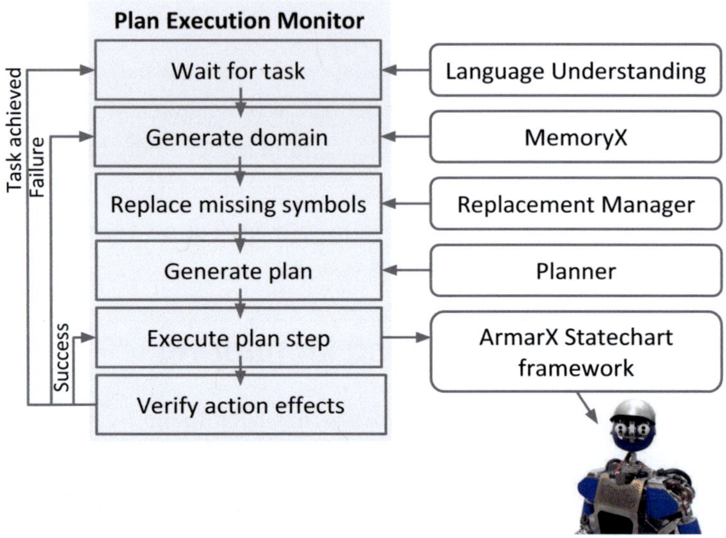

Figure 6.7: Control flow of the task execution and monitoring with the main components for each phase.

If some effects could not be found in the current world state or if the execution of the action failed, the planning component is called again to generate a new plan with the same goal, but on the current world state instead of the initial world state. If the special failure result *object not found* occurs, the symbol replacement manager is called beforehand to generate a new location hypothesis or to replace the object with a new object.

Only effects relevant to the current action are checked. Changes of another agent to the scene or unexpected effects of the executed action which do not contradict the effects of the current action are tolerated by the execution monitoring and the robot continues unless the changes of another agent's action directly violate the execution of the current solution. If the robot

completes the plan successfully, it waits for further instructions. In the case of failure, replacing is performed as long as a potential solution theoretically exists, or planning is terminated, e.g. via human instruction.
To be more flexible, the task execution approach supports multiple execution modes. The default mode is described above and takes a goal state description as input. Another mode is the *direct command* mode. In this mode, it is possible to directly evoke single planning operators such as *grasp* or *open*. Yet a full parametrization of the action is required here to be given to the robot, which is computed by the planner in the *planning* mode. The *direct command* mode is used in (Kaiser et al., 2016) in a pilot interface for semi-autonomous execution of actions suggested based on the affordances of geometric primitives. The action parametrization is here generated based on the perception of the robot and the suggestion selection of the user.

6.8 Summary

In this chapter, an approach for execution of tasks in dynamic environments on a humanoid robot was presented. To reason about the current state of the dynamic environment, the robot's perception is converted online into a symbolic domain description, which is used by a symbolic planning system to generate an action sequence for a given task. Additionally, the domain description is altered automatically based on multi-modal replacement strategies if the domain contains unusable objects or if locations for objects are missing. These strategies use information such as common sense knowledge about object affordances or object locations learned from text mining, shared visual features regarding the affordances of objects and common places of objects learned from experience. The generated action sequence consists of actions that correspond to the robot skills presented in the previous chapter. A task monitoring keeps track of the execution and triggers correcting measures if necessary.

7 Evaluation

In this chapter, the developed methods are evaluated. First, the hierarchical segmentation and action recognition approach is evaluated. This segmentation is compared to two other segmentation methods known from the literature. Second, several use cases for the statechart concept are presented and discussed. Finally, the task execution approach is shown in a complex scenario and evaluated on the robot ARMAR-III and in a user study.

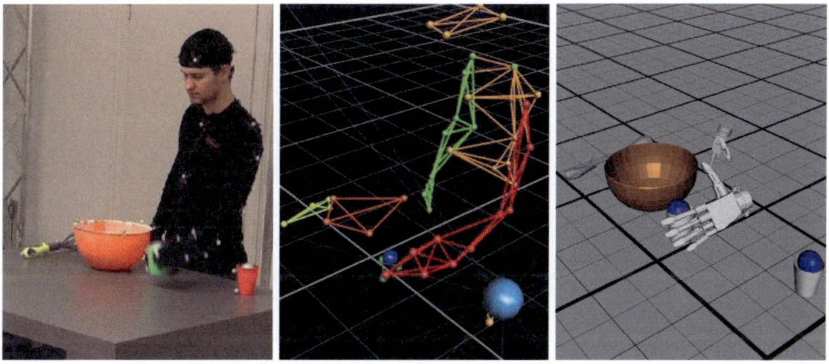

Figure 7.1: Stages of demonstration capturing and processing: The human demonstrator with attached markers on the objects (left); marker group representation (middle); 3D mesh models with applied 6D object pose (right). *Source:* (Wächter and Asfour, 2015) © 2015 IEEE

7.1 Task Segmentation and Action Recognition

This section starts with the employed experimental setup for capturing human demonstrations and a discussion about the quality of the preprocessing step, converting marker positions into 6D object poses. The segmentation results of the presented approach are compared with a manual reference segmentation and other segmentation methods based on a new segmentation metric. The action recognition algorithm is evaluated by comparison to a labeled dataset.

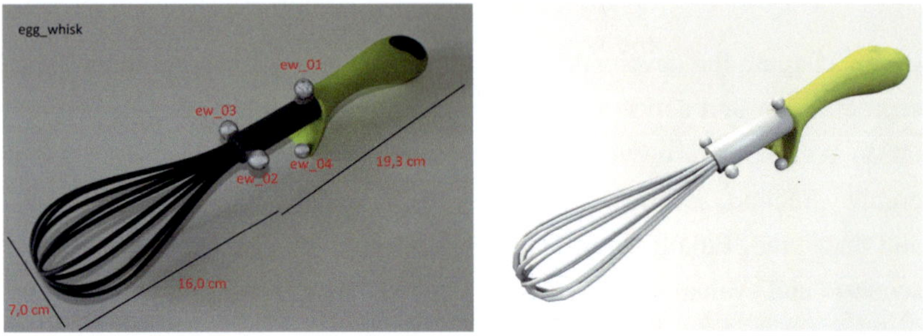

Figure 7.2: *Left*: Image of a real object with attached reflective markers from the KIT Whole-Body Human Motion Database (Mandery et al., 2016b). *Right*: A visualization of the corresponding 3D mesh model with attached virtual markers.

7.1.1 Experimental Setup

Human demonstrations were recorded with the marker-based VICON motion capture system. The capturing system consists of ten cameras for the observation of the scene, in which all objects are common rigid household objects that have at least three markers attached to them in an asymmetric arrangement. The human demonstrator is, depending on the dataset, fully captured or only her/his hands are captured. All demonstrations are available in the KIT Whole-Body Human Motion Database[1] (Mandery et al., 2016b). Nonetheless, regarding the human motion, only the hands are used for the segmentation and action recognition. The motion capture system uses models of the spatial marker relations of all objects for automatic labeling of

[1] `https://motion-database.humanoids.kit.edu`

the markers (see Figure 7.1, middle). For every object, a 3D mesh model is available, created either with a 3D scanner or by hand and enriched with virtual marker positions (see Figure 7.2).

7.1.2 Segmentation Metric

To compare the presented approach to other algorithms, a metric is proposed that measures the error of the key frame extraction in square seconds compared to a reference segmentation. The metric is similar to the F1-score (van Rijsbergen, 1979), but considers additionally the accuracy of the true positives. This metric represents the precision of the results better since the temporal precision is not only binary. Let K_r be the set of key frames of the reference segmentation and K_f the found key frames of the algorithms. The metric assigns each $k_r \in K_r$ the closest key frame available in K_f. Each key frame can only be assigned once and the squared error to the reference key frame is measured. The maximum allowed distance of a correct key frame to the reference key frame was chosen to be one second; otherwise the key frame is considered *missed*. For every missed key frame (false negative) and for every false positive key frame of the algorithms, a penalty is added, in order to severely penalize completely wrong key frames. The penalty factor was chosen by the author to appropriately reflect the missed and unmatched key frames since there exists no objective penalty factor. In summary, the metric e used for comparison is given by:

$$e = (m + f) \cdot p + \sum_i \min_j (k_{r,i} - k_{f,j})^2, \tag{7.1}$$

where m is the number of missed key frames, f the number of false positives, k_r and k_f are the reference key frame and corresponding detected key frame.

7.1.3 Evaluation Methodology

The presented Hierarchical Segmentation (HS) approach needs to be compared to an ideal segmentation to assess the quality of the results. Though, generating automatically ground truth segmentation data of demonstrations is not feasible since the actions often change seamlessly into the next action. Additionally, even for a human it is not easy to determine when an action ends and the next action starts. Nonetheless, the recordings of these action sequences were segmented manually as a reference to which the results of the algorithms are compared. The presented segmentation approach and its sub-segmentation methods are compared to the segmentation method Zero-Velocity-Crossings (ZVC) and to a method based on Principal Component Analysis (PCA) (see subsection 3.1.2). All parameters of all methods were optimized over all demonstrations with random cross-validation to achieve the best average results with one set of parameters. The PCA and ZVC methods were optimized manually, while the Hierarchical Segmentation (HS) was optimized with a genetic algorithm due to the higher number of hyperparameters.

7.1.4 Experiments and Datasets

To test the presented segmentation approach the following task demonstrations were recorded among others: *preparing dough*, *wiping a table*, *shaking/pouring of a bottle*, *drinking from a cup*, *cutting with a knife* and *polishing a bowl*. These tasks were chosen to have a wide range of typical manipulations of a kitchen environment and are similar to the demonstrations used in other publications (Aksoy et al., 2011; Ramirez-Amaro et al., 2014). Since the two levels of the hierarchical segmentation target different types of actions, the recordings were split into two datasets to show the benefit of the different segmentation levels. Eventually, both sets were merged and evaluated together.

For each task, between two and five repetitions of the tasks with variations in the action selection, action duration and order of the actions were recorded. In total, 30 task demonstrations are used for the evaluation. Figure 7.3 shows snapshots of the demonstrated tasks.

Dataset 1

Dataset 1 predominantly contains actions with observable object relation changes such as grasping. The datasets consists of 17 demonstrations with multiple actions, i.e. segments. In total, the dataset contains 67 segments of the reference segmentation. The actions in the dataset are mostly grasping and placing actions, but also cutting, pouring, mixing, wiping, drinking actions.

The *preparing dough* task is in this dataset and the task setup contains a table, a cup, a bottle, a bowl and a whisk (three bottom rows of Figure 7.3). The cup is grasped by the human, the content is poured into a bowl and then the cup is placed again on the table. This is repeated for the bottle. Afterwards the liquids are mixed with a whisk, which also has to be grasped by the human to perform the mixing action. In the end, the whisk is placed again on the table. To note here is that the liquids are not tracked by the motion capture system. This task is chosen because it contains several objects and typical actions in the context of a household robot.

The other demonstrations are short demonstrations mostly consisting of a tool use action from the list described previously, for which the tool needs to be grasped and placed.

Dataset 2

Dataset 2 predominantly contains actions without observable object relation changes such as shaking and wiping. The datasets consists of 13 demonstrations with multiple actions, i.e. segments. In total, the dataset contains 291 segments of the reference segmentation. In the following, some of the tasks contained in the dataset are described.

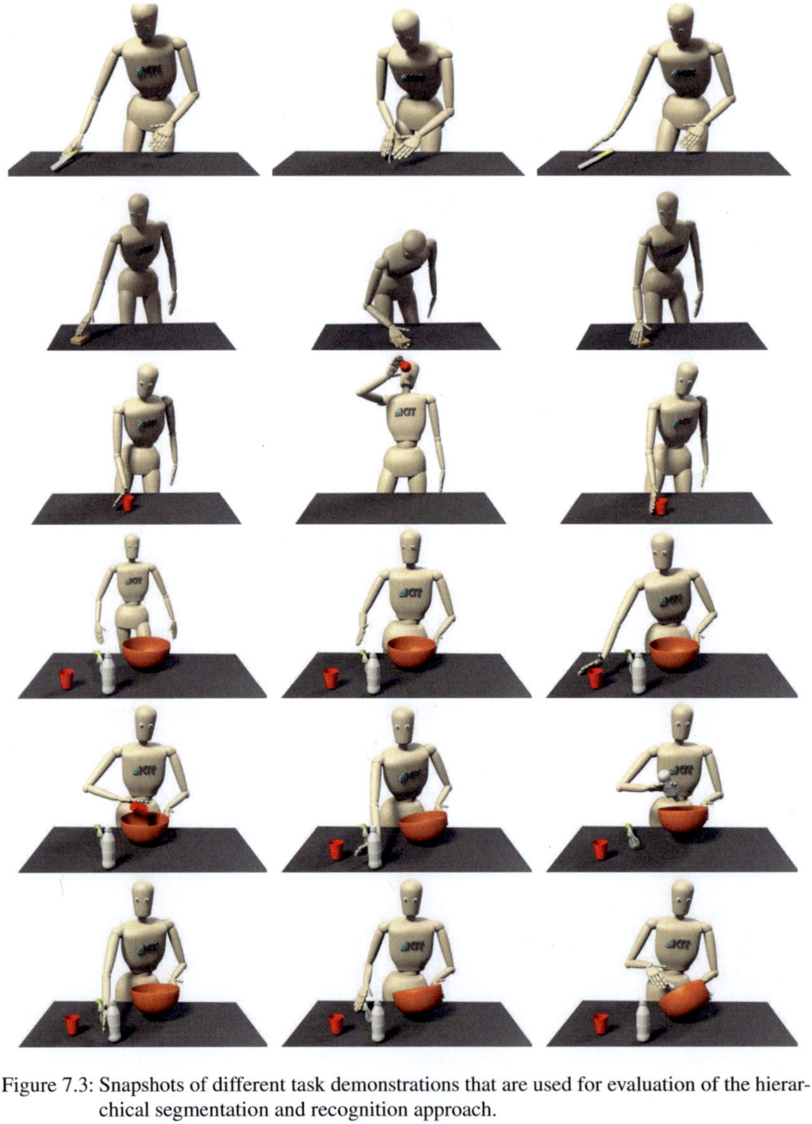

Figure 7.3: Snapshots of different task demonstrations that are used for evaluation of the hierarchical segmentation and recognition approach.

In the *table wiping* task, the human demonstrator grasps a sponge from a table and wipes the table using several different wiping styles like intensive wiping of a spot or wiping of a wide area with circle motions. In a second task, a bottle is being grasped, tossed, inspected, shaken, poured and dripped off. In a third task, a big bowl is being held in one hand and polished by the other hand with changes of the hands in between.

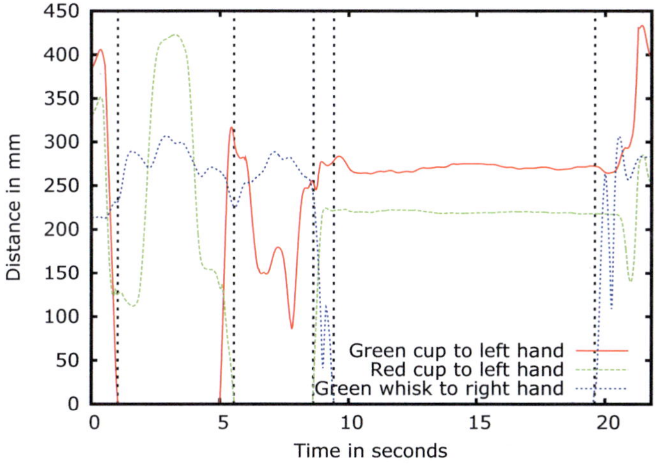

Figure 7.4: Segmentation based on object contact relations: When two objects get in contact with each other (contact in this case is approximated as *distance* < 7 mm) or lose contact, a new key frame is inserted with the current world state attached to it. The dotted vertical lines depict the detected semantic segments. Only distances between objects that get in contact during the complete demonstration are shown. *Source:* (Wächter and Asfour, 2015) © 2015 IEEE

7.1.5 Segmentation based on Semantics

Since the two levels of the hierarchical segmentation are independent segmentation algorithms, it makes sense to assess them individually and compare them with the hierarchical segmentation. An example of the semantic segmentation is shown in Figure 7.4 related to the *preparing dough* task showing picking and placing of a cup and a whisk and stirring with the whisk. The inter-object distances that are not relevant for this segmentation task are omitted in this diagram for better comprehensibility. Whenever the left or right hand grasps an object or puts down an object a new key frame is inserted. Table 7.2 shows the segmentation results on a dataset consisting of actions with visible effect, i.e. contact relation changes, such as most actions shown in Figure 7.3. This segmentation method proves to be reliable for actions with visible effect, which are predominantly manipulation actions. The accuracy is better and almost all key frames also appear in the reference segmentation than of the motion segmentation and of ZVC and PCA as can be seen in Table 7.2, Table 7.3 and Table 7.4 (rows *accuracy* and *unmatched key frames*). Yet due to the nature of the method not all key frames can be found, which is reflected in the row *missed key frames* especially in *Dataset 2*. The parametrization used is shown in the top part of Table 7.1 at the end of this section. The distance threshold of 6.81 mm allows for precise detection of contact-relations with a tolerance for model discrepancies or capturing errors. The semantic merge threshold determines when effects are considered to result from the same action and thus also determines the minimum segment length. With 47 frames, i.e. 0.47 seconds, the merging still allows short actions with natural execution speed.

7.1.6 Segmentation based on Motion Characteristics

An example for the segmentation based on motion characteristics is shown in Figure 7.5. It can be seen that the algorithm inserts key frames whenever the motion characteristic noticeably changes, e.g. change between periodic and discrete motions, different amplitudes of a periodic motions, different frequency etc. Table 7.4 shows that the motion characteristics segmentation yields already good results in the range of the hierarchical segmentation and significantly better than ZVC and PCA. Yet Table 7.3 shows that discrete motions as in *dataset 1* are not the strength of this segmentation algorithm, but it still achieves comparable results. As for all motion based segmentation approaches, the number of false positives (unmatched key frames) is higher than that of the semantic segmentation since the motion data is more ambiguous than contact relations. Also the accuracy is not as good as the accuracy of the semantic segmentation since seamless actions do not always have a clear segmentation point in motion space. However, the results on the combined dataset (Table 7.4) are better than the results of the semantic segmentation alone since the motion characteristics based segmentation has the chance to find all key frames. And indeed, it extracts most of the actions. The used parameters are shown in Table 7.1.

Table 7.1: Parameters of the hierarchical segmentation algorithm (see section 4.3 and 4.4) and their values.

	Parameter description	**Value**
Semantic Segmentation	Distance threshold (ε)	6.81 mm
	Semantic merge threshold	47 frames
	Hysterese factor (λ)	1.72
	Velocity difference (τ) between objects	1 mm/frame
Motion Segmentation	Window size (w)	0.54 sec
	Minimum segment size (l_{min})	1.09 sec
	Minimum segment quality (μ)	818
	Normalization weight (z)	3.78
	Gauss filter width (σ)	0.078

The minimum segment size is bigger here (1.09 seconds) since when the motion style changes there are several possibilities where to insert the key frame, i.e. where the condition holds that left and right of the key frame the motion characteristics are different. The sliding window size is 0.54 seconds wide, which means half a second of motion is considered for the comparison. Since even marker-based motion capture data contains noise, a Gaussian filter with small width has been applied to reduce acceleration spikes, which would have direct effect on the segmentation quality since it is an acceleration based heuristic.

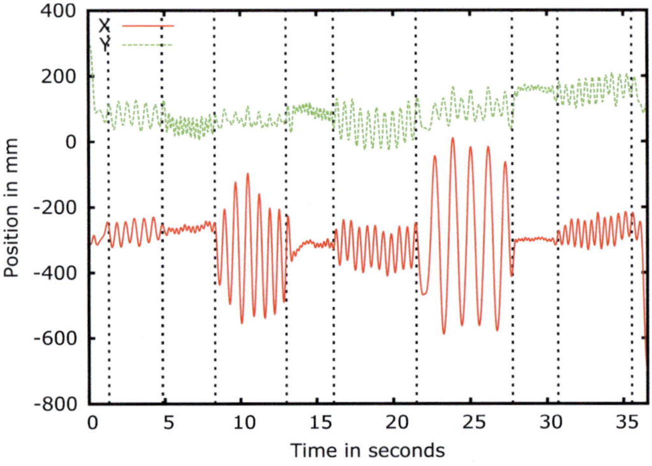

Figure 7.5: Sub-segmentation of one semantic segment (sponge touches hand and table) into different subsegments, i.e. different wiping styles. The dotted vertical lines depict the segmentation points. Whenever the wiping styles changes, a key frame is inserted. *Source:* (Wächter and Asfour, 2015) © 2015 IEEE

7.1.7 Hierarchical Segmentation

In the previous sections, each level of the hierarchical segmentation was evaluated individually to show the benefit of combining both segmentation methods into one hierarchical segmentation. Now, the semantic segmentation approach is used for the top level and the motion characteristic based segmentation is used for the sub-segmentation for the bottom level of the hierarchical segmentation as described in chapter 4.

Table 7.2: Comparison of the hierarchical segmentation to the two levels used individually with *Dataset 1*. HS = Hierarchical Segmentation, SS = Semantic Segmentation, MCS = Motion Characteristics Segmentation

Average Results	**HS**	**SS**	**MCS**
Error	**2.89 s^2**	**3.14 s^2**	**3.97 s^2**
Accuracy	0.22 *s*	0.21 *s*	0.33 *s*
Unmatched key frames	0.54	0.0	0.18
Missed key frames	1.9	2.0	1.72

Table 7.3: Comparison of the hierarchical segmentation to the two levels used individually with *Dataset 2*.

Average Results	**HS**	**SS**	**MCS**
Error	**2.94 s^2**	**5.43 s^2**	**3.03 s^2**
Accuracy	0.25 *s*	0.18 *s*	0.25 *s*
Unmatched key frames	2.0	0.23	1.55
Missed key frames	3.4	10.0	4.04

Figure 7.6 shows the *shaking-pouring* task and the results of the compared algorithms. The semantic level of the hierarchical segmentation extracts three segments from the demonstration (green bars in Figure 7.6). These segments are in proximity of the key frames of the reference segmentation and correspond to picking up the bottle, putting it back on the table and retreating the hand from the bottle. The algorithm found all actions with an observable effect. Yet the found segments contain actions without observable effect. Furthermore, the transitions between the segments are seamless, which increases the difficulty to extract them. The motion characteristic level found several subsegments (red bars in Figure 7.6). Six of them are fairly close to key frames of the reference segmentation and four do not correspond to any key frames. Though, the performance on this difficult demonstration is not

perfect, the presented approach outperforms the other methods. The PCA based segmentation oversegments the demonstration due to the complexity of the motion. ZVC shows better performance, but also extracts too many false positive key frames. The results show that a maximum complexity per segment as a segmentation criteria, as PCA does, is not a suitable criteria to extract actions out of a demonstration. Motions can be very complex and still be part of the same action. Furthermore, the change of direction on multiple dimensions of the motion, as used by ZVC, is also not a suitable criteria, especially for periodic motions, which change the direction in every period.
Figure 7.7 shows results of two further demonstrations in detail: *wiping* with a sponge and another trial of the *shaking-pouring* task. The wiping task contains fast, periodic movements without stopping. It can be seen that ZVC barely finds segmentation points since it is mostly a continuous motion whereas PCA oversegments the demonstration due to the complexity of the motion trajectory.
Figure 7.8 shows the repeatability of the approach on different trials of the same task. The *shaking-pouring* task is demonstrated three times with different order and duration of the actions. The quality of the segmentation result is comparable in all three trials. Only zero or one false positive key frames are extracted by the proposed method. Nevertheless, some key frames from the reference segmentation are missed (two to five key frames). This is due to marginal changes in the motion characteristic between the segments. All semantic segments are found.

Average results over all demonstrations

The average metric results over all 30 task demonstrations are shown in Table 7.4. The Hierarchical Segmentation (HS) algorithm is compared with the segmentation algorithms based on ZVC and PCA using the proposed metric. To apply the metric to all 30 demonstrations (15 - 40 seconds each) the results of the algorithms are compared to the manual reference segmen-

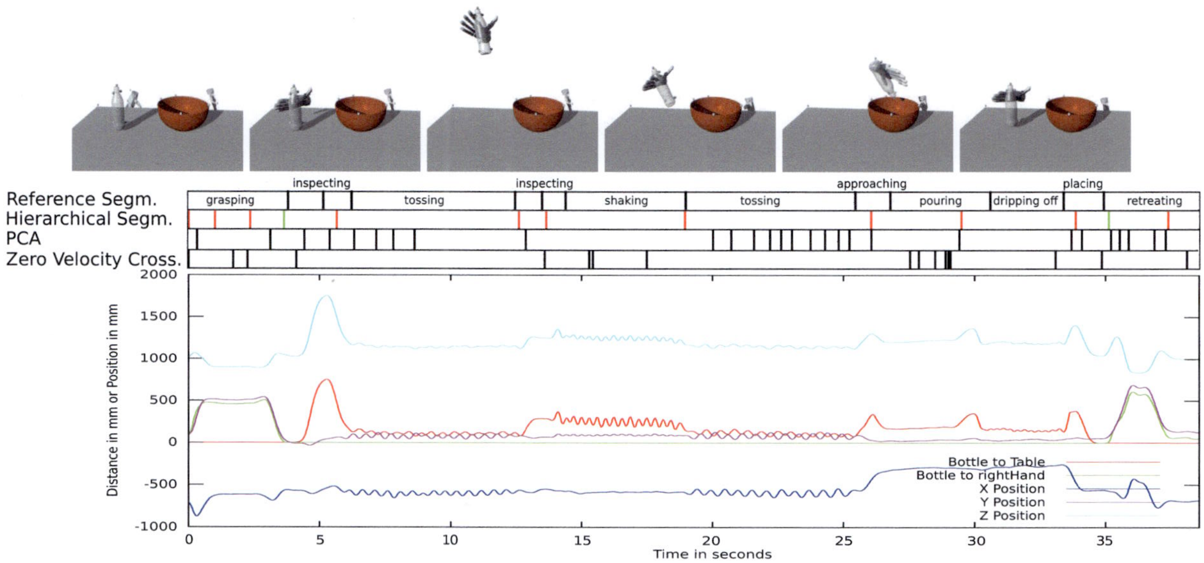

Figure 7.6: Comparison of the HS, PCA and ZVC with a manual reference segmentation. Key frames are indicated by vertical lines.

The red lines in the Hierarchical Segmentation denote the key frames found by the motion characteristic segmentation and the green lines denote the semantic key frames. The curves show the position of the right hand in Cartesian coordinates and relevant distances between objects during the task execution.

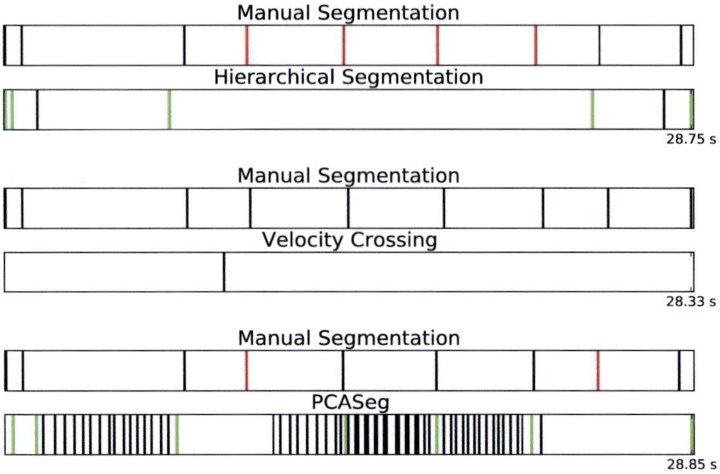

(a) Segmentation of a *wiping* task

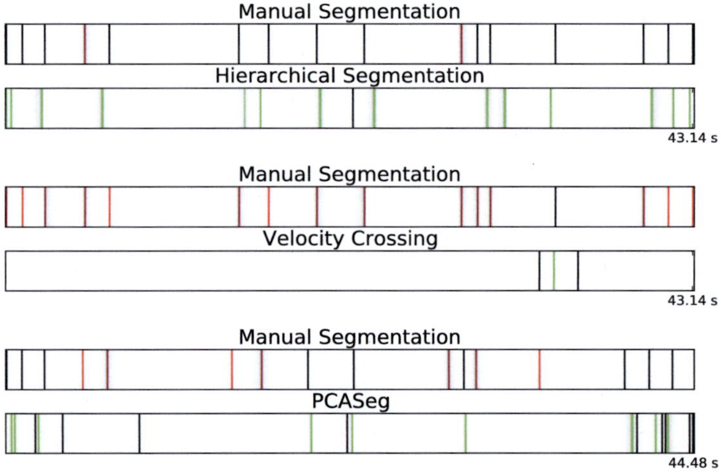

(b) Segmentation of a *shaking-pouring* task

Figure 7.7: Comparison of the proposed approach to Zero-Velocity-Crossings and PCA-based segmentation on the (a) *wiping* task and the (b) *shaking-pouring* task Missed key frames are shown in red. Matched key frames are shown in green. The different lengths of the plots originate from the last key frame found in the segmentation results.

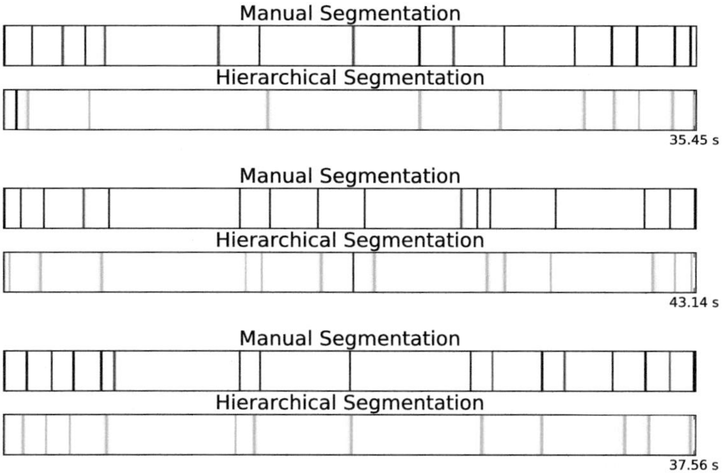

Figure 7.8: Comparison of manual to hierarchical segmentation with three trials of the same *shaking-pouring* task. The vertical lines denote key frames at their timestamps. This shows the repeatability of the approach on variant yet similar trials. Missed key frames are shown in red. Matched key frames are shown in green. The three trials contain grasping, placing, shaking, tossing, inspecting and pouring actions similar to Figure 7.6 (top) each in different order and with different timing. *Source:* (Wächter and Asfour, 2015) © 2015 IEEE

Table 7.4: Comparison of average results over 30 task demonstrations with other segmentation methods based on the metric proposed in subsection 7.1.2. The *error* value is the metric value. The other criteria are part of the *error* value.

Average Results	**HS**	**SS**	**MCS**	**ZVC**	**PCA**
Error	**2.93 s^2**	**5.1 s^2**	**3.14 s^2**	**7.01 s^2**	**20.18 s^2**
Accuracy	0.24 s	0.19 s	0.26 s	0.1 s	0.36 s
Unmatched key frames	1.51	0.15	1.09	0.6	27.9
Missed key frames	2.9	7.33	3.27	12.27	3.5

tations created each by two persons. The *error* row shows the value of the proposed metric. The other rows are the sub-results of the metric and make up the final metric value as described in subsection 7.1.2. *Accuracy* is the

average distance of matched key frames to the corresponding key frames of the reference segmentation. In other words, it shows how close an extracted key frame lies to the reference segmentation if a match for this key frame was found. *Unmatched key frames* (false positives) mean how many key frames found by the algorithms were not assigned to a key frame in the manual segmentation. *Missed key frames* (false negatives) indicate how many key frames of the manual segmentation per demonstration where not assigned to a key frame found by the algorithms. The results are the average results over the 30 task demonstrations. The parameters for the presented approach were trained on five task demonstrations with a genetic algorithm and tested on all 30 demonstrations. It can be seen that the proposed algorithm achieves significantly smaller error values than the two other approaches. The *accuracy* and *unmatched key frames* values in the case of ZVC are better since the number of key frames inserted by ZVC is considerably smaller (compare *missed key frames* and see the examples in Figure 7.6 and Figure 7.7) and therefore the probability of inserting a false positive is decreased.
PCA on the other hand has a comparable amount of *missed key frames*, but a significantly higher number of *unmatched key frames*.

7.1.8 Psychological Study: Comparison to Segmentation by Human Subjects

The hierarchical segmentation algorithm was also used in a collaboration with the university of Groningen. They conducted a study (Schlichting et al., 2018) on the segmentation of demonstrations and reproduction of the duration of actions by humans after watching tasks containing several actions. Twenty psychology students with no affiliation to robotics and no knowledge about how the algorithms work were asked to extract the start- and end-points of actions in videos shown to them. Figure 7.9 shows the results of the segmentation experiment. For each action the average start and end points of the actions of all participants are shown together with the variance.

The black star denotes the result of the hierarchical segmentation. It can be seen that the hierarchical segmentation produces very similar results to the segmentation by the participants and therefore indicates that the meaning of the segments found by the algorithm is similar to how humans segment sequences of actions.

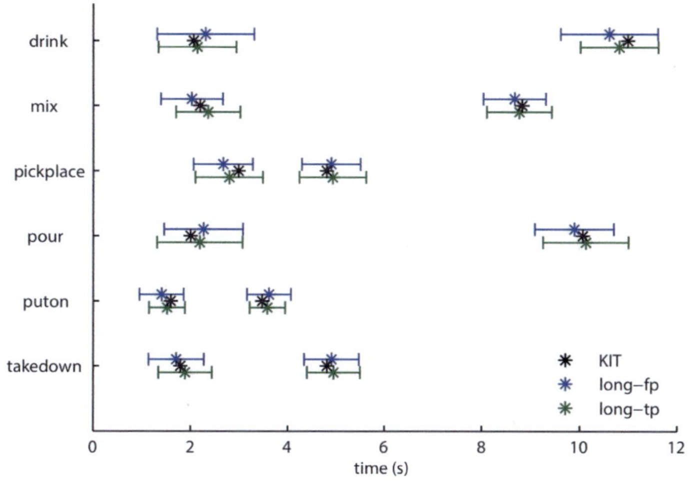

Figure 7.9: Psychological study on segmentation of demonstrations. *Source:* (Schlichting et al., 2018) © 2018

7.1.9 Action Recognition

The action recognition presented in section 4.5 was evaluated on the same dataset using a manual segmentation of the demonstration, in which each segment was enriched with multiple labels of different abstraction, such as *graspBottle, grasp, discrete*. In total, the dataset contains 31 different actions such as *approach* or *pour* and 105 executions of these. This means there are only on average 3.39 repetitions per action available, which increases the difficulty for the identification of the relevant action features. The decision tree classifier was evaluated with the cross-validation technique by selecting

a random training and test dataset 100 times with 90/10 split and averaging the results over all iterations.
Most actions are labeled with multiple labels with different levels of abstraction. Yet the most abstract description still corresponds to the abstraction of a planning operator such as *pour*. The more concrete labels describe the style or the object the action was performed on, e.g. *xwiping*, meaning wiping along the x-axis, or *approachBottle*. These more concrete labels are very difficult to recognize since they describe the motion in detail and also have very few examples in the dataset.
The average recognition success on the dataset is 76.2 %. A successful recognition is achieved if one of the labels was predicted successfully. Table 7.5 shows the complete action recognition results for each action label. Both, the recognition of the action with semantic effects works well (e.g. *approach*, *place*) and the abstraction from the used object was successful. The recognition of the actions without semantic effect does not succeed as well since the used dataset is very challenging in this regard. Some of the actions do have very similar motions such as shaking and tossing a bottle and therefore a mismatch is easily possible. For example, the *wiping* action has a recognition rate of 97.58 % since there are no actions with similar motion characteristics using the sponge in the dataset, though the motion characteristics for *tossing* are very similar. Wrong recognitions between actions with different objects did not happen (e.g. wiping with a sponge and tossing a bottle). Therefore, it can be said that the distinction between actions based on the semantic state could be achieved.
The recognition results for the more concrete action labels are considerably worse (e.g. *ywiping* or *approachBottle*) since the motions between different concrete labels are very similar or only have very few examples in the dataset. Yet the concrete labels are not needed for association with planning operators. They were used to test the limits of the algorithm and would only be needed for parametrization of the action style.

Table 7.5: Action recognition results for each action label.

Action labels	Recognition rate in %	Action labels	Recognition rate in %
approach	83.72	*pouringBottle*	32.00
approachBottle	88.46	*raise*	29.41
approachSponge	64.71	*raiseBottle*	29.41
drippingoff	47.62	*retreat*	100.00
drippingoffBottle	47.62	*retreatBottle*	93.33
inspect	37.93	*retreatSponge*	100.00
inspectBottle	37.93	*shaking*	47.22
lift	92.00	*shakingBottle*	47.22
liftBottle	92.00	*spotwiping*	75.56
lower	57.14	*tossing*	65.91
lowerBottle	57.14	*tossingBottle*	65.91
place	100.00	*wiping*	97.58
placeBottle	100.00	*xwiping*	81.25
pour	81.48	*xywiping*	52.24
pourBottle	81.48	*ywiping*	58.82
pouring	32.00	**Average**	**76.20**

7.1.10 Discussion

The segmentation approach has proven to find segments that are similar to segments chosen by humans. But first, the segmentation sub-algorithms will be discussed separately.

Semantic segmentation

In general, the semantic segmentation approach relies strongly on the precision of the 6D trajectory and the model associated with it. An inaccurate model can lead to false or missed contact detection and therefore to false segmentation. Based on experience, the precision is sufficient to detect all contacts and segment the trajectory into their semantic parts. As shown in Figure 7.4, where the action sequence of pouring with two different cups and a subsequent mixing action was demonstrated, the contact between the objects can be extracted from the distance curve and, thus, the key frames are detected. In certain cases, a loss of contact does not mean that a new action starts. For example, during the wiping action, the sponge occasionally loses contact with the table. Based on the assumption that actions have a minimum duration (470 ms is used here), this situation is avoided to a certain degree by merging adjacent key frames that are temporally too close to each other. Although the semantic segmentation performs well for most demonstrations, the drawback of the semantic segmentation becomes apparent if actions occur without any observable effects, for which the semantic segmentation has no chance to detect these actions.

Segmentation based on Motion Characteristic

The sub-segmentation tackles this problem of unobservable effects of actions. Additionally, different styles of periodic actions can be detected (different wiping styles, e.g. wiping in lines or intensive wiping on one spot). In Figure 7.5, a segmented action of wiping is shown, which is then subsegmented in different wiping styles. Each time the wiping pattern changes, a new key frame is inserted. In Figure 7.6, the further inspection and segmentation of a semantic segment, which supposedly represents a pouring action, indicates that these segments comprise more actions without observable effects, and, thus, can be further divided into subsegments. In the experiments, most segments have been detected, though a smooth transition between two actions can be problematic and no key frame might be found.

Hierarchical Segmentation

Table 7.2 and 7.3 highlight the benefit of combining both segmentation algorithms to a hierarchical segmentation. The semantic segmentation performs well on *dataset 1* whereas the motion characteristic based segmentation performs well on *dataset 2*. The hierarchical segmentation outperforms both algorithms on both datasets. Both algorithms have their strengths and weaknesses, which can be well compensated with the combination of both. The motion characteristic based segmentation achieves good results, but can be improved with the hierarchical segmentation. Yet the semantic segmentation, and thus the hierarchical segmentation, provides besides the segmentation the valuable information about the semantic state at each segment, which a motion based segmentation cannot provide. While this information is not important for the segmentation itself it is particularly valuable for further processing of the segmentation results such as action recognition or learning of effects of actions (Aksoy et al., 2015).

Action Recognition

The semantic state is used in the action recognition approach. In contrast to the state of the art, the proposed approach uses motion and semantic information to recognize actions. The combination of both feature spaces allows to reliably distinguish actions with similar motions (*tossing* vs. *wiping*). Such actions are typically difficult to distinguish only based on motion features. Yet not all motions can be recognized. Motions without semantic change and no characteristic motion are not reliably recognized. For example, the pouring motion is basically an idle motion with the bottle in the hand since the bottle is just held over the container. However, most other approaches would probably also fail to recognize such action. Due to the global nature of the motion feature descriptor and the search space reduction induced by the semantic state, the proposed approach is able to recognize actions with high variance on trajectory level.

Since the employed feature vector is high dimensional, it is important to detect relevant dimensions. The problem is that the relevant dimensions are action dependent and, thus, general dimension reduction cannot be applied. The decision tree classifier is able to identify many irrelevant dimensions and can recognize irrelevant differences in the demonstrations. For example, in some demonstrations the hand touches the table at the end of the *approach*

action. This contact is not relevant for this action, which is successfully recognized by the classifier.
As for all data-driven approaches, more recognition results would have helped the classifier to create a better decision tree since many different actions are to be recognized, where each action type naturally increases the ambiguity and the need for a bigger dataset. A comparison to other approaches is unfortunately hardly possible since other approaches either use motion features or semantic information. Furthermore, a public dataset providing the needed information for such a comparison is also not available.

7.2 Statecharts

In the following, the use of statecharts for the development and programming of robot capabilities is described. Several use cases, the developed generic robot skills, example applications as well as fault recovery capabilities are presented.

7.2.1 Robustness and Fault Recovery

Due to the distributed nature of the statecharts, it is possible to recover from fatal faults (i.e. process crashes) of sub-statecharts. The statecharts are organized in groups, which typically contain semantically or functionally similar states. Since each statechart group resides in its own process, fatal faults of other groups do not result in a complete system shut down. If a substate of another group crashes or terminates due to internal errors, the parent state will receive a failure transition from the substate and can handle the failure. This is an important feature for a complex robot like a humanoid robot with several subsystems, especially bipedal humanoid robots, where a crash of a single substate could have serious consequences for the whole system resulting e.g. in a fall of the robot which can be dangerous to humans and to the robot itself. Therefore, a powerful fault recovery is an integral part of a robot that interacts with humans.

7.2.2 Generic Robot Skills

Reusability of developed robot skills is an important requirement. Considering the reusability of software in robotics is still not prevalent, maybe due to the fact that the research field is relatively young and the research efforts focus mainly on solving robotic problems while ignoring aspects of software engineering for robotic applications. The developed statechart approach addresses important aspects of software reusability. e.g. by utilizing a profile hierarchy to specify robot specific configurations where they are needed (see subsection 5.3.7). This significantly supports the development of robot independent skills, which can be resused on other systems.

Table 7.6: Subset of available generic robot skills implemented with statecharts.

Skill	**Description**
JointControl	Synchronous joint control
CartesianControl	Cartesian TCP control
HolonomicPlatformControl	Controls a holonomic platform
GazeControl	Controls the view direction in Cartesian space
PositionVisualServoing	Visually tracked, Cartesian TCP control
GraspObject	Grasps an object from a location
ZeroForce	Enables zero force control on an end effector
PlaceObject	Places an object on a surface
ScanForObject	Scans the environment for objects
ShapeHand	Moves a robot hand into a specific shape
BringObject	Brings an object to the human
PlanExecution	Handles the execution of planning steps

A subset of the most important skills, which have been developed, implemented and evaluated is given in Table 7.6. These skills range from low-level controllers like *JointControl* and *HolonomicPlatformControl* to high-level functionalities statecharts like *BringObject* and *PlanExecution*, which handles the execution of generated plans on the statechart level. In total, over 300 states have been implemented for the robots ARMAR-III and ARMAR-4.

7.2.3 Use Cases

In the following, example use cases for different applications of statecharts are described. The *BringObject* statechart is a realization of a complex task whereas the *Walking* statechart shows an example of state-based coordination. The third use case *Execution of Planning Operators* illustrates an advanced example, in which statecharts are used for the execution of Object-Action Complexes as the underlying elements of a symbolic plan.

Use Case: *BringObject* Statechart

The Figures 7.10 to 7.14 show the first five levels of the *BringObject* task statechart. With this statechart it is possible to specify fetch and deliver an object to a human. Each of the statechart levels is dedicated to a different abstraction level of the task and could be used again in another statechart for the realization of another robot capability. The parametrization is passed on from the top-level to the lower-level where and when necessary. Green states are remote states, located in another statechart group process and yellow states are final states, which trigger an event in the parent state.

Several states like *StopRobot*, *MoveJoints*, *ShapeHand* or *VisualServoTowardsObject* are used multiple times in this task, which illustrates the reusability of states in different situations. Figure 7.15 shows the input parameters of the *GraspObject* state. Most of the parameters are pre-configured for each robot type based on simulation or expertiments on the real robot. Some of the parameters (e.g. *ObjectName*) do not have any default values and need to be mapped from the parent or previous state, when the state is used in another state. The types vary from basic types like `float` and `string` to complex types like `Vector3` or `ChannelRef`, which is a reference to a dynamic data channel of an ArmarX observer, e.g. an object instance channel of the *ObjectMemoryObserver*[2].

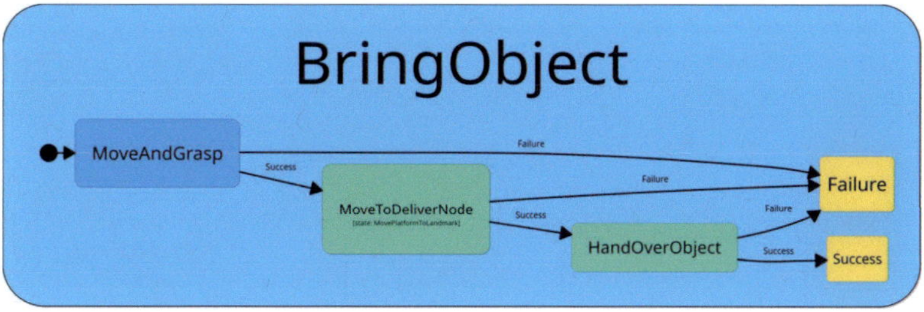

Figure 7.10: Top level of the *BringObject* task statechart, in which the robot moves to a given location, grasps an object and hands it over to a human at a delivery location. This and the following statecharts are exported directly using the graphical statechart editor.

This state is not designed for one specific robot, but for a robot type with certain features: It needs at least one arm with a gripper or hand, knowledge about the current object location and a mobile, holonomic platform. To run this statechart on a new robot only a robot model, which is given in the Simox (Vahrenkamp et al., 2012) file format, and an adaption of the robot specific statechart parameters (e.g. the kinematic chains) is needed.
This *BringObject* statechart already represents a complex task and consists of several states, which are used as planning operators in the task execution system (see subsection 7.3.2) (i.e. the states *GraspObject, MovePlatform-ToLandmark, HandOverObject*).

Bipedal Walking Coordination with Statecharts

The statechart concept is also used to implement bipedal walking on the humanoid robot ARMAR-4 (Asfour et al., 2013). Since walking consists of repeated phases or in other words repeated states it makes sense to use statecharts for the coordination of the walking controllers.

[2] The *ObjectMemoryObserver* contains all the object localization results.

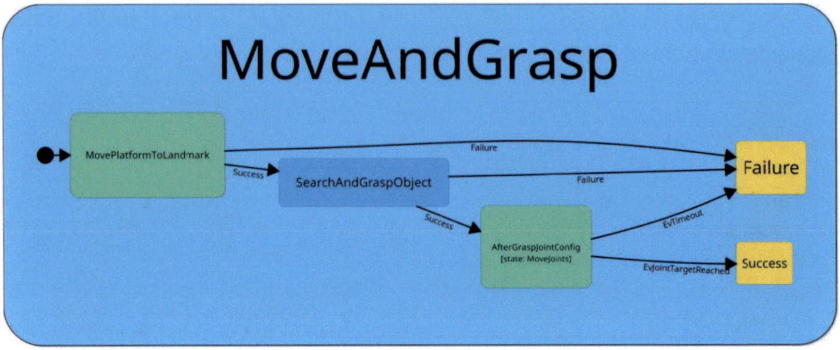

Figure 7.11: Second level of the *BringObject* task statechart: move to a location and grasp an object.

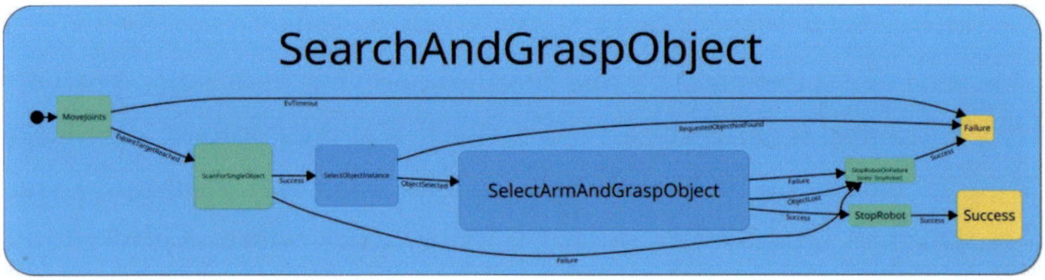

Figure 7.12: Third level of the *BringObject* task statechart: visual object localization and grasping.

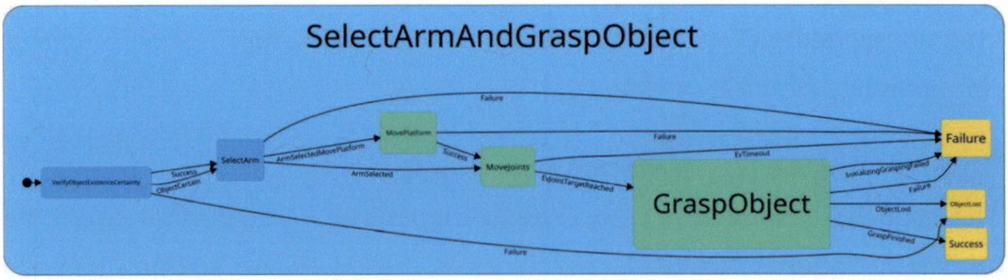

Figure 7.13: Fourth level of the *BringObject* task statechart: select an arm, reposition the mobile platform and grasp the object.

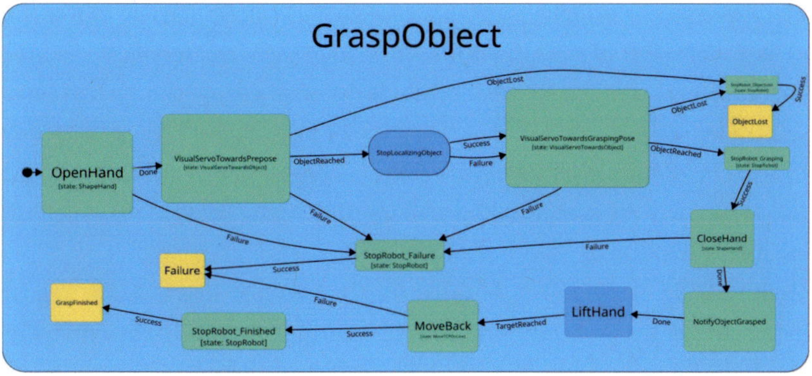

Figure 7.14: Fifth level of the *BringObject* task statechart: Execute grasping using visual servoing.

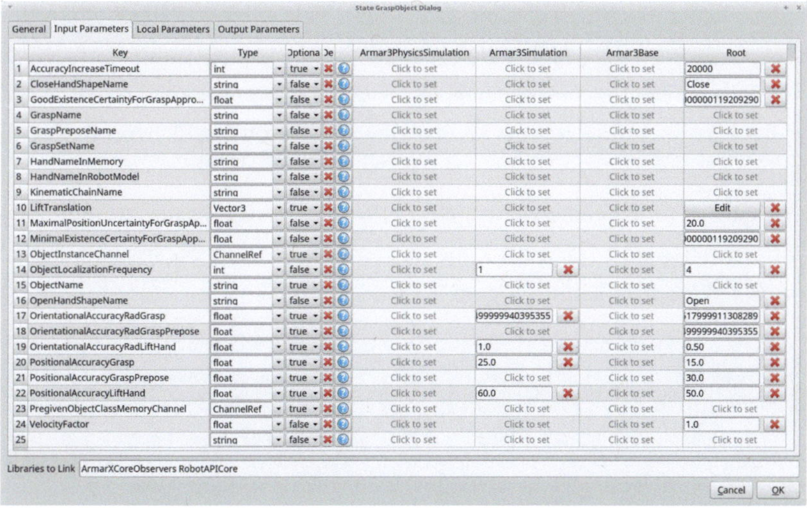

State GraspObject Dialog

General | Input Parameters | Local Parameters | Output Parameters

	Key	Type	Optiona	De	Armar3PhysicsSimulation	Armar3Simulation	Armar3Base	Root
1	AccuracyIncreaseTimeout	int	true		Click to set	Click to set	Click to set	20000
2	CloseHandShapeName	string	false		Click to set	Click to set	Click to set	Close
3	GoodExistenceCertaintyForGraspAppro...	float	false		Click to set	Click to set	Click to set	)00000119209290
4	GraspName	string	false		Click to set	Click to set	Click to set	Click to set
5	GraspPreposeName	string	false		Click to set	Click to set	Click to set	Click to set
6	GraspSetName	string	false		Click to set	Click to set	Click to set	Click to set
7	HandNameInMemory	string	false		Click to set	Click to set	Click to set	Click to set
8	HandNameInRobotModel	string	false		Click to set	Click to set	Click to set	Click to set
9	KinematicChainName	string	false		Click to set	Click to set	Click to set	Click to set
10	LiftTranslation	Vector3	true		Click to set	Click to set	Click to set	Edit
11	MaximalPositionUncertaintyForGraspAp...	float	false		Click to set	Click to set	Click to set	20.0
12	MinimalExistenceCertaintyForGraspApp...	float	false		Click to set	Click to set	Click to set	)00000119209290
13	ObjectInstanceChannel	ChannelRef	true		Click to set	Click to set	Click to set	Click to set
14	ObjectLocalizationFrequency	int	false		Click to set	1	Click to set	4
15	ObjectName	string	true		Click to set	Click to set	Click to set	Click to set
16	OpenHandShapeName	string	false		Click to set	Click to set	Click to set	Open
17	OrientationalAccuracyRadGrasp	float	true		Click to set	)99999940395355	Click to set	;17999911308289
18	OrientationalAccuracyRadGraspPrepose	float	true		Click to set	Click to set	Click to set	)99999940395355
19	OrientationalAccuracyRadLiftHand	float	true		Click to set	1.0	Click to set	0.50
20	PositionalAccuracyGrasp	float	true		Click to set	25.0	Click to set	15.0
21	PositionalAccuracyGraspPrepose	float	true		Click to set	Click to set	Click to set	30.0
22	PositionalAccuracyLiftHand	float	true		Click to set	60.0	Click to set	50.0
23	PregivenObjectClassMemoryChannel	ChannelRef	true		Click to set	Click to set	Click to set	Click to set
24	VelocityFactor	float	false		Click to set	Click to set	Click to set	1.0
25		string	false		Click to set	Click to set	Click to set	Click to set

Libraries to Link: ArmarXCoreObservers RobotAPICore

Cancel | OK

Figure 7.15: Input parameters of the *GraspObject* state.

The implemented walking algorithm combines 1) calculation and movement of the robot center of mass (CoM) based on joint angle measurements, inertial measurements and a robot model with its dynamic parameters such as mass and center of mass, with 2) the utilization of force measurements acquired by a 6D force-torque sensors in the ankle to compensate model errors. In the following, only the aspects of the walking algorithm related to the statechart are briefly explained.

Each walking step consists of three phases, i.e three states, which can be parametrized for each leg. The corresponding statechart is shown in Figure 7.16. Since walking requires high-performance controllers, not all of them are implemented in the state user code. The ankle-stabilizer controller, e.g. uses only the force-torque information to minimize torques and to keep the robot stable. Therefore, the controller is executed close to the hardware to reduce latency between sensing and acting. This ankle-stabilizer controller offers a network interface which is used for its parametrization and activation. The parametrization depends on the current step phase and is performed in the corresponding state as soon as the state is entered. Performing such state-dependent parametrization in sequential code would mean to represent the states in sequential code, which results in unreadable code. Additionally, with the statechart framework the current state and its parameters can be inspected graphically (see Figure 7.18).

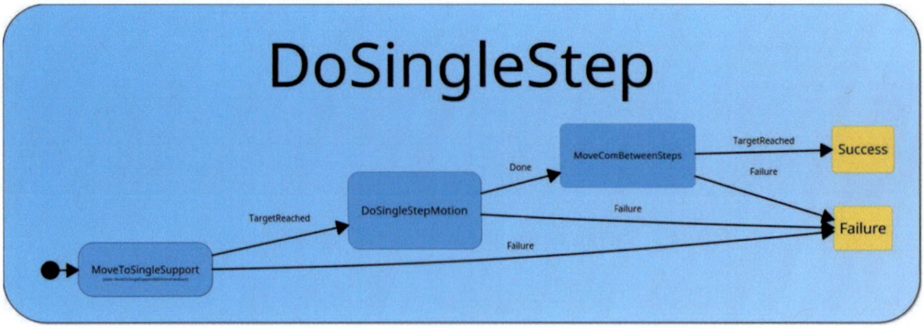

Figure 7.16: State for a single step with the bipedal robot ARMAR-4.

Since the low-level controllers should not execute long-running calculations, the inverse kinematics solving for the center of mass movement is performed asynchronously in the *running* phase of the state, which guarantees that the state is still responsive to sudden state changes. To perform continuous steps, the state *DoSingleStep* is instantiated three times with different parametrization: initial step with half step size, right step and left step. The right and left step states are structured in a loop to perform an arbitrary numver of steps in succession (see Figure 7.17).

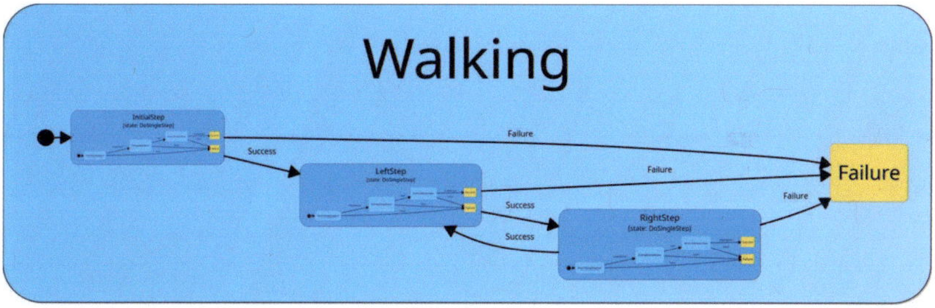

Figure 7.17: Walking statechart with three instantiations of a single step with different parametrization.

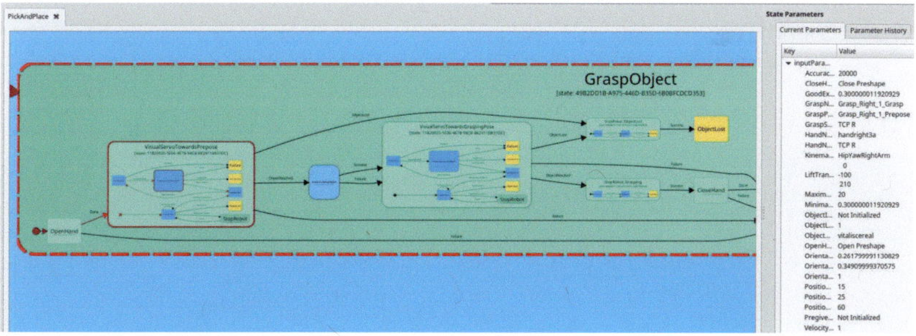

Figure 7.18: Each running statechart can be inspected online with the StatechartViewer GUI, including the currently active states (red border), the state trace (red transitions) and the state parametrization.

Execution of Planning Operators

The third use case is the execution of plans generated by the approach presented in chapter 6. The fact that statecharts usually have a static layout and plans are sequences of parametrized planning operators, which depend on the current world state and goal, allows the assumption that the statecharts cannot fulfill the requirements for executing symbolic plans. To deal with this important aspect, the *DynamicRemoteState* was developed (see subsection 5.3.10). This dynamic remote state can morph into any remote state at runtime upon entering the state and, thus into the desired planning operator state.

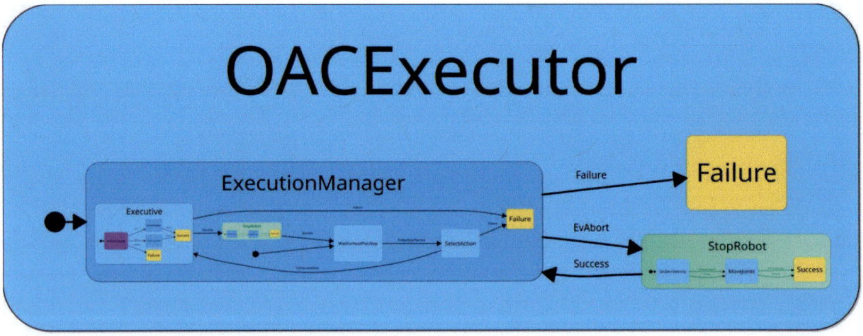

Figure 7.19: The top level state for OACs execution encapsulates the *ExecutionManager* state to allow unexpected abort of the exection (event *EvAbort*).

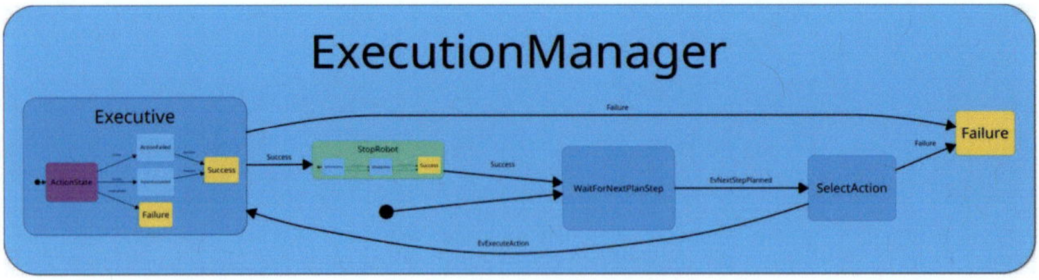

Figure 7.20: Substate of the OAC execution that manages the selection and execution.

Figure 7.19 and 7.20 show the statechart implementing the logic for executing smybolic plans. The *OACExecutor* state is merely an encapsulating state which defines a clean procedure for a unexpected abort of the execution, e.g. when a `stop` command triggered by the human is recognized. In such situations, the robot is stopped by the `EvAbort` and returns then back to the *ExecutionManager* state, in which it waits for a new command. The *ExecutionManager* and *Executive* state (Figure 7.20) perform the actual work. The *WaitForNextPlanStep* substate observes a remote data field that is controlled by the plan execution component and reacts whenever this data field is updated. This data field contains a plan step with the corresponding the parameters. This plan step is given to the *SelectAction* substate, in which the associated OAC is queried from the long-term memory. The OAC contains the information about the associated statechart, which is passed on to the *Executive* substate. The purple colored *ActionState* is a *DynamicRemoteState*, which morphs into the selected state and executes it. After finishing, either with success or some failure, the result is sent back to the plan execution component to react appropriately, the statistical measure of the OAC is updated and the statechart waits for the next step of the plan.
With this statechart it is possible to execute any planning operator, i.e. OAC, that has an entry in the OAC segment of the long-term memory, without altering this statechart. This statechart has been used e.g. in (Kaiser et al., 2016) as element of a graphical pilot interface for semi-autonomous action selection and execution. The interface suggests a variety of applicable OACs and automatically parametrizes theses OACs based on the perceived environment. A parametrized OAC is then send to the proposed task execution system (see chapter 6) for execution.

7.2.4 Discussion

Up today, more than 300 states for a wide variety of applications for the robots of the ARMAR series have already been developed and extensively been used not only to demonstrate simple tasks, but also for complex skills of humanoid robots acting in real world scenarios, including grasping, mobile manipulation, learning from human observation and the execution of symbolic plans.

The decision to remove some of the features of Harel's original statechart for the proposed statechart approach has proved to be a right decision, since the removed features (inter-level-transitions, history-connector) were rarely missed in robotic applications and their removal has led, from robotics point of view, to significant improvements regarding comprehension, reusability and maintenance.

Compared to the MCA-framework (Scholl et al., 2001) that was used before for the ARMAR robots, in which robot behaviors were also encoded by state machines, but without a defined structure, the proposed statechart concept improved the development and maintenance of robot skills for humanoid robot. In particular, the explicit definition and specification of the data flow in the statecharts have contributed to reusability, convenient debugging and inspection of the system at different levels of abstraction.

An important effect of the explicit data flow definition is that implicit data-dependencies to other states are not possible, which ensures that entering a state with the same set of input-parameters leads to the same result. On the one hand, specifying the data flow explicitly and in great detail can be seen as an overhead in the development, but on the other hand the easy and convenient debugging and better maintenance will certainly make the effort worth in the long run. Additionally, specifying and inspecting data flow with graphical tools simplifies this process considerably. Such graphical tools are not only useful for defining the data flow, but indispensable for developing complex state machines. Hence, the graphical statechart editor is one of the most important tools of the ArmarX framework.

The necessity seen by Harel for introducing the concept of hierarchies into state machines can be confirmed. Hierarchies are essential for developing robot skills as compositions of elementary sub-skills and for maintaining the

reusability. The grasping skill, for example, consists of up to six hierarchy levels, where some of the substates are used several times. Unrolling this into one hierarchy level results into a statechart that is practically impossible to design by hand due to the number of required states.
The proposed statechart concept proved to be applicable for use cases on all levels of a robot architecture. An example of a low-level statechart is a controller for holonomic platform movements, where the leaf state is the PD-controller (using the asynchronous user code *run*-function for high performance control) and the level above decides on the waypoints. An example for a high-level statechart is the statechart for OAC execution presented earlier (see Figure 7.20).

7.3 Task Solving and Execution in Dynamic Environments

The third part of this thesis, the task solving and execution in dynamic environments is evaluated in a complex scenario, in which the robot, in cooperation with a human, has to solve the task of rearranging a room and preparing a meal. The scenario was executed on the humanoid robot ARMAR-III and in a user study in simulation to test the usability and robustness of the approach with untrained users. The scenario, the experiments and the results are presented in the following.

7.3.1 Predicate Providers

Predicate providers map subsymbolic information into symbolic predicates and are essential for the execution of tasks in dynamic environments. The reliability and precision of the predicate providers have a high impact on the robustness of the execution and failure handling. Yet the reliability of predicate providers depends on the quality of the sensor data, e.g. of perceived object poses and the location of the robot in the environment and thus such data should be pre-processed to reduce noise and task specific errors. In the case of a mobile manipulation task, a Kalman filter is used for the fusion of the visually determined object pose and for the robot pose estimation based on odometry and laser scanner information.

All predicate providers developed in this thesis are rule-based and use the results of the self-localization, object localization and robot configuration. Table 7.7 shows the list of used predicates that are extracted from robot perception, and an explanation of the calculation rule. As described in section 6.7 there are more predicates that are unobservable by the robot and tracked based on the effects of actions. In the use case described in the following sections, the predicates proved to be of varying reliability. For example, the `left/rightGraspable` predicate is reliable since it is represented by bounding boxes in which objects are rarely on the borders. Even if objects are on borders, they can be graspable by the other hand leading to a successful execution of the task without even noticing a failure. The `inHand` predicate is less reliable since it depends on the predicted object pose in the hand, a challenging task which is not supported by the visual object localization method applied in this thesis.

Table 7.7: Observed predicates used in the planning domain.

Predicate with types	**Description**	**Calculation description**
inHand(object, hand, robot)	*object* is in *hand* of *robot*.	Distance between object and hand $< \varepsilon_1$
objectAt(object, location)	*object* is at landmark *location*.	Distance between object and location $< \varepsilon_2$
robotAt(robot, location)	*robot* is at landmark *location*.	Distance between robot and location $< \varepsilon_3$
handEmpty(robot, hand)	*hand* of *robot* is empty.	Distance of hand to all objects $> \varepsilon_4$
leftGraspable(object)	Left hand grasp is known for *object* and it is currently at a suitable location for the left hand.	Object position in bounding box
rightGraspable(object)	Right hand grasp is known for *object* and it is currently at a suitable location for the right hand.	Object position in bounding box

7.3.2 Evaluation Scenario

The task execution system has been evaluated on the humanoid robot ARMAR-III (Asfour et al., 2006) in a kitchen environment. ARMAR-III is a 43 DOF humanoid robot with two seven DOF arms, two hands with five fingers each, a seven DOF head with two stereo camera systems with foveal and peripheral lenses and a holonomic mobile platform with three omni wheels. It can operate in an autonomous way using only onboard sensors, computing power, batteries.

The case study is a dinner preparation scenario, in which the robot and a human cooperatively prepare a salad and rearrange the room for dinner for two people. Figure 7.21 shows snapshots of the execution in chronological execution order. The video resulting from the *Xperience* (Xperience, 2011) project which demonstrates the results is online available[3].

During the dinner preparation, the human instructs the robot to perform certain tasks using natural language[4] while the human is performing other parts of the dinner preparation. To solve the tasks the task execution system is employed and several robot skills described by the developed statechart representation are used. Table 7.8 shows the list of symbolic skills used in this scenario.

The sub-tasks of the scenario are:

1. Fetching and placing the salad bowl on the sideboard
2. Preparing the salad with corn and oil
3. Placing the salad bowl on the dining table
4. Fetching a chair and moving it to the dining table
5. Delivering a beverage to the human

[3] `https://youtu.be/-8oC-WW5P1I`

[4] The natural language understanding component is not part of this thesis.

Each of the sub-tasks are uttered to the robot, which tries to find a solution based on the current world state. Each of the tasks could also be commanded with direct action commands without using the planning system. In this mode, the full parametrization needs to be given to the robot, e.g. *grasp the green cup with your left hand from the sideboard*. The execution can be stopped at any time and the robot awaits a new command for solving a new task based on the current world state.

Table 7.8: Symbolic robot skills used in the evaluation scenario.

Skill	**Description**
JointControl	Synchronous joint control
CartesianControl	Cartesian TCP control
HolonomicPlatformControl	Control of a holonomic platform
GazeControl	Control of the view direction in Cartesian space
PositionVisualServoing	Visually tracked, Cartesian TCP control
GraspObject	Grasp an object from a location
ZeroForce	Enable zero force control on an end effector
PlaceObject	Place an object on a surface
ScanForObject	Scan the environment for objects
ShapeHand	Move a robot hand into a specific shape
BringObject	Bring an object to the human
PlanExecution	Handle the execution of planning steps

7.3.3 Natural Human-Robot Interaction User Study

The presented task execution approach was evaluated together with a language understanding component developed in (Ovchinnikova et al., 2015; Wächter et al., 2018) in a user study. The goal of the study was to evaluate whether the approach can handle commands of untrained users and achieve

(a) Grasping the bowl

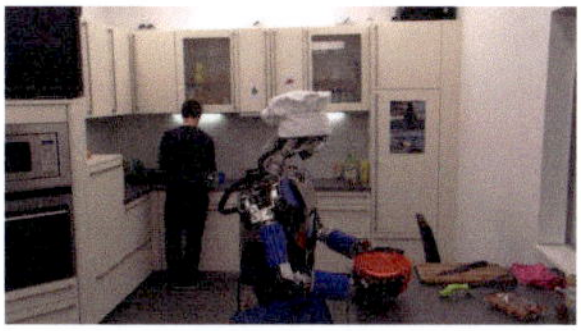

(b) Placing the bowl

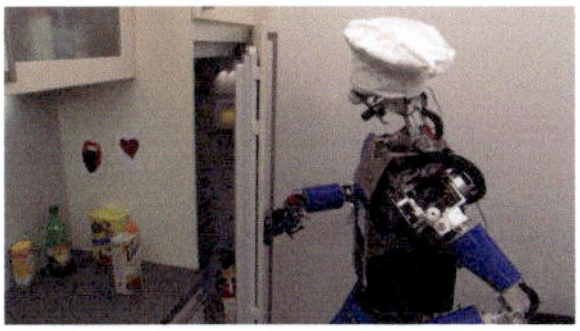

(c) Opening the fridge

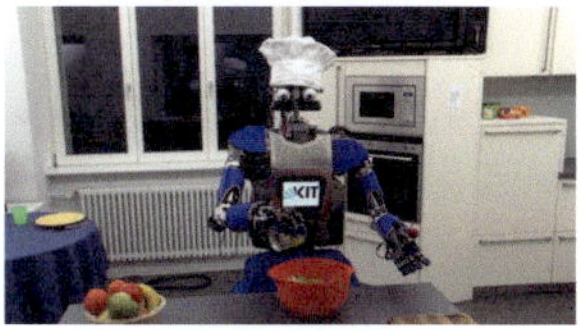

(d) Pouring corn

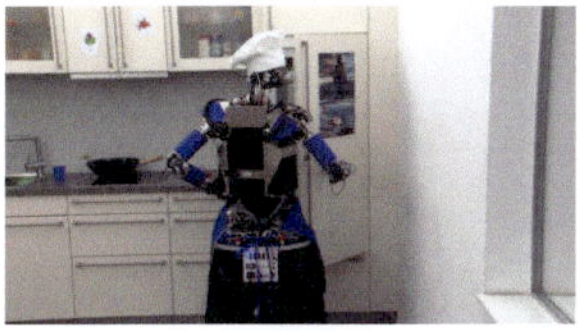

(e) Closing the fridge

(f) Pouring oil

(g) Stirring salad

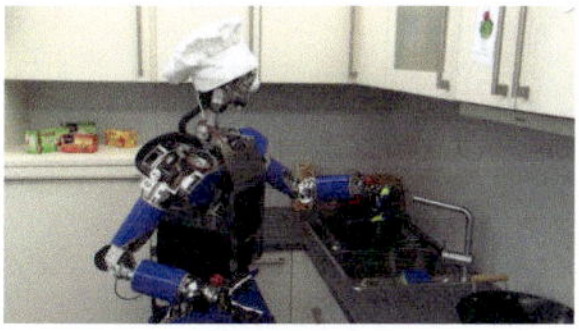

(h) Tidying up

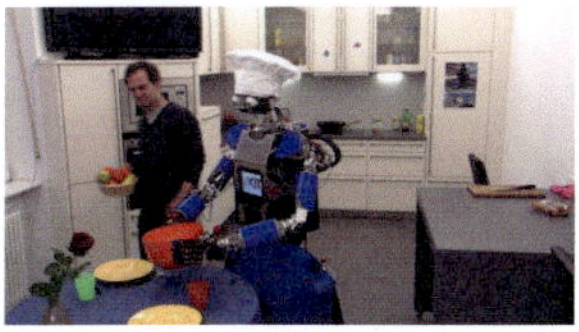

(i) Placing the bowl

(j) Bringing juice

Figure 7.21: Snapshots of cooperatively rearranging the room and preparing dinner. *Source:* (Wächter et al., 2018) © Elsevier 2018

the desired goal. Due to the use of generic interfaces, a simulation and the real robot are interchangeable without changes of the execution system. For this experiment a kinematic simulation is used and the user can follow the robot in a 3D view of the current working memory content of the robot. This graphical user interface is shown in Figure 7.22. It visualizes the current state of the working memory and allows the user to enter commands in natural language as text.
The subjects were instructed to achieve a given goal by controlling the robot with natural language commands in the simulation environment. The setup of the study is defined as follows:

- The user should communicate with the robot by typing or speaking sentences via the GUI.
- The initial world state with labels of robot locations and ambiguous object names, as shown in Figure 7.23(a), was presented to the user.
- The goal state, which the user should achieve by controlling the robot as shown in Figure 7.23(b).
- The drink on the table could be chosen from a predefined list of drinks represented as images (e.g. beer, juice, milk) and the salad should contain ingredients, which were also only represented as images of corn and oil.

The experiment was conducted in the lab with no time constraints. In total, eight subjects (four male, four female, age 20–60) with no background in robotics or related areas took part in the experiment. The experimenter was present during the experiment to transcribe the process, but did not give any further instructions and answered no questions. An example of two sentences with one failure and one success is shown in Table 7.9.

Table 7.9: Experiment transcript example

User ID	Command	RM	Memory Update	Planner	Execution	User feedback
Al	"Put two cups on the table"	-	-	Failed: No plan, since only one location `table` known	-	User confused, since no feedback
	"Set the table for two people"	-	-	OK	OK	-

The command *Put two cups on the table* resulted in a failure since there was only one location of type `table` available, but the planner tried to put both cups on that one location. This results in a loop since the planner wants to put one cup on the table and removes it in the next step again to put the other cup on the table. A functioning command for this task would be *Set the table for two people*. This command maps the location type `placesetting` to two different instances of that type and resulting in a quickly found plan.

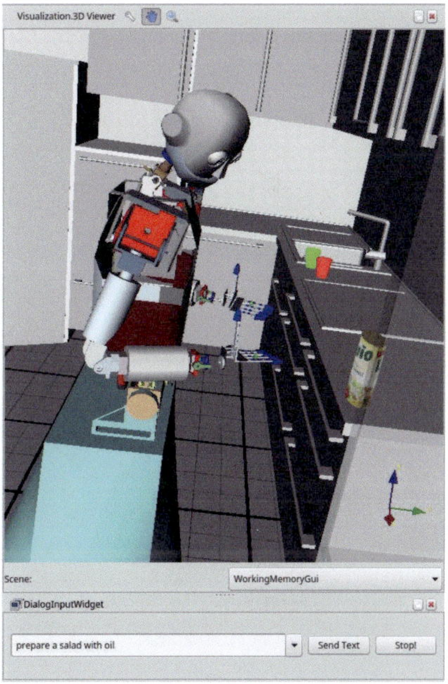

Figure 7.22: Graphical user interface for interaction for interaction with the simulated robot.

All subjects were able to solve the tasks using 12.5 utterances in average. The commands given by subjects differed greatly. Some were using general task descriptions, e.g. *make a salad with corn and oil*, whereas others were giving fine granular commands such as *go to the fridge. Open the fridge. Take the corn.*

In general, most commands could be handled by the system. Few commands failed due to various reasons, such as an invalid mapping of natural language instructions to the planning goal.

The Replacement Manager (RM) suggested six object replacements, of which four were accepted by the users. It has also generated location hypotheses for all known objects and asked the users for unknown object locations. The working memory was successfully updated during the execution and after processing human utterances. Given the generated domain description and a goal that could be fulfilled, the planner generates the required plans to achieve the goal.

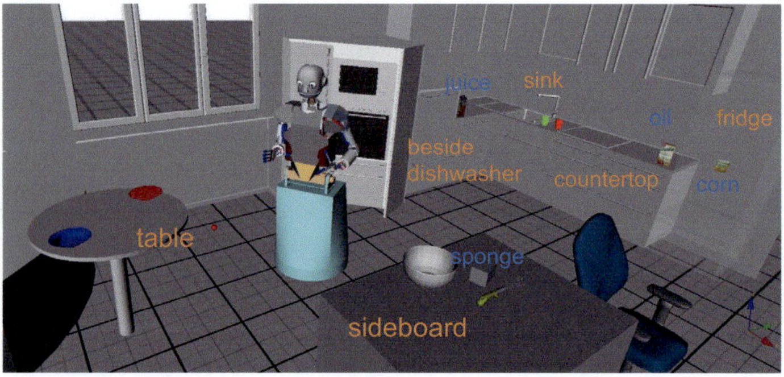

(a) Initial scene

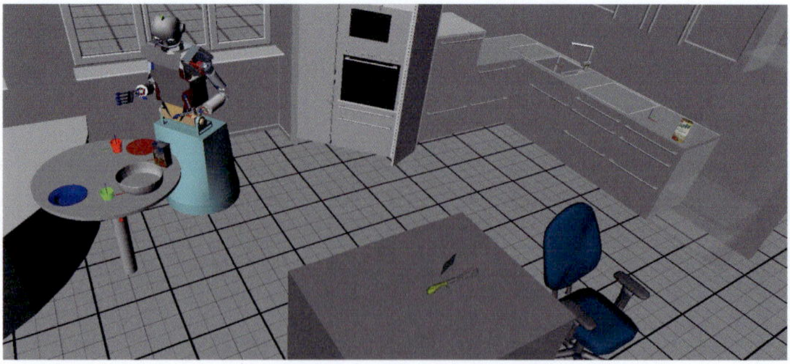

(b) Goal scene (table setup)

Figure 7.23: Initial and goal scenes provided to the users.

An important issue that occurred during the experiments were cases in which several replacements of the same type were required in a goal expression. For example, the command *Put two glasses on the table* implies that there should be two different glasses on the table. Since no glasses are available in domain, the RM suggests replacing them with cups and selects the type `red_cup` as a replacement. Since there is only one instance of `red_cup`, the new goal implying two different instances could not be fulfilled. To tackle this problem, the RM should take the number of instances of replaceable types into account and should provide only replacement suggestions if the required instances are available.
The lack of a mechanism for estimating the interdependence between consecutive goals has led to unreasonable sequences of plans. For example, a user has instructed the robot to put a cup on the table, which the robot could perform without problems. Some of the follow-up commands may result in a plan that instructs the robot to remove the cup again from the table, which may contradict the original intention of the user. Yet this problem is difficult to solve since sometimes previous goals are not relevant for future plans, e.g. putting the bowl first on the sideboard to prepare salad and placing it later on the dining table would contradict the first goal, but is still correct.

Table 7.10: Runtime of the main components in seconds. *Source:* (Wächter et al., 2018) © Elsevier 2018

Component	**Average**	**Maximum**	**Minimum**
Language Understanding	0.26 s	0.79 s	0.13 s
Visual Features Strategy	2.61 s	4.61 s	1.66 s
Planning System	0.89 s	7.99 s	0.13 s
Action Execution	20.59 s	42.01 s	4.44 s

The runtime of the main components is depicted in Table 7.10. The host PC used for the measurements was an Intel Core i7-6700HQ CPU running at 2.60 GHz with 16 GB RAM. The tasks from the user study were used to measure the runtime. The Language Understanding component performs its analysis reliably in under one second (maximum 0.79 s). No input data produced a noticeable delay in the task triggering. The RM and its strategies except the visual features strategy are table look ups of precomputed information, e.g. common sense object replacements, and consume no significant CPU time. These strategies were omitted in Table 7.10. The visual features strategy on the other hand is computationally expensive and required 2.61 s on average. The runtime mainly depends on how many point cloud clusters (i.e. objects) have been found in the current scene. The runtime of the planning component highly depends on the specific goal and the current state of the memory and varies between 0.15 ms and 7.99 s for the given tasks. The minimum of 0.15 ms was achieved when the goal was already fulfilled before planning. The maximum of 7.99 s was required for the task *set the table for two people*, with all objects except the corn already being in the working memory. The runtime of the action execution was measured per plan step in the simulation. This measurement includes the full duration of the robot's action execution. On the real robot the runtime is similar, but slightly higher due to smaller joint angle velocities and due to a higher perception time. The runtime highly depends on the executed action; in case of the *move* action it depends on the travelling distance.

7.3.4 Discussion

With the presented task solving and execution approach it is possible to equip a robot with the powerful capability to act and interact autonomously in an unstructured and dynamic environment. The ability to reason about the location and the potential replacement possibilities of objects enhances the autonomy and flexibility of a robot. The versatility of the location hypothesis and object replacements is ensured due to the multi-modality of the symbol replacement, which can easily be extended by additional strategies in the future. However, the location hypothesis and object replacements are not and cannot always be correct, since there is no way to know if an object is currently at the most probable location. A wrong replacement will be discovered during execution and the robot will autonomously correct it by creating a new hypothesis. Apart from the common places strategy, this will be always the same hypotheses since the underlying data structure is static. One possible extension would be to assess the reliability of hypotheses after execution and adjust the confidence of the employed strategies and individual hypotheses. This way, the robot could ground the hypotheses in the real world and improve their quality over time.

The predicate providers allow a developer to easily design a mapping between continuous subsymbolic and discrete symbolic representation needed for reasoning and planning. These need to be developed for each observable predicate used in the planning domain. Optimally, all predicates should be observable to always allow a grounding of symbolic information in the real world, but this is not feasible with current state of the art sensors and perception methods. In this thesis, the most important and perceivable predicates about the object and robot location are automatically extracted from sensor data. Additional predicates such as *graspable* are derived from information stored in the robot's memory. Information about non-observable predicates can be provided to the robot via natural language, such as *the fridge is closed*. From theoretical point of view, each of the predicates could be declared *non-observable*, but then online domain generation and task execution monitoring would be pointless since the execution would be independent from any information about the current state of the world. Thus, the correctness of the generated planning domains relies mainly on the quality of these predicate providers. In future work, predicate providers that are learned by the robot itself based on simulation or real world experience would enable the robot to learn new symbolic information and possibly refine existing predicate providers. Similarly, to the extension of the replacement strategies, the predicate providers could analyze the execution results that are executed based on the information from the predicate providers and adapt those in failure situations.

The user study showed that the proposed approach can be controlled by untrained users to solve a complex task with a humanoid robot. Problems that the participants encountered showed that the feedback of the robot, i.e the human-robot-interaction, should be improved. The robot should be able to communicate the reason of failure and externalize its current state to the participants to realize more symbiotic interaction and reliable task execution.

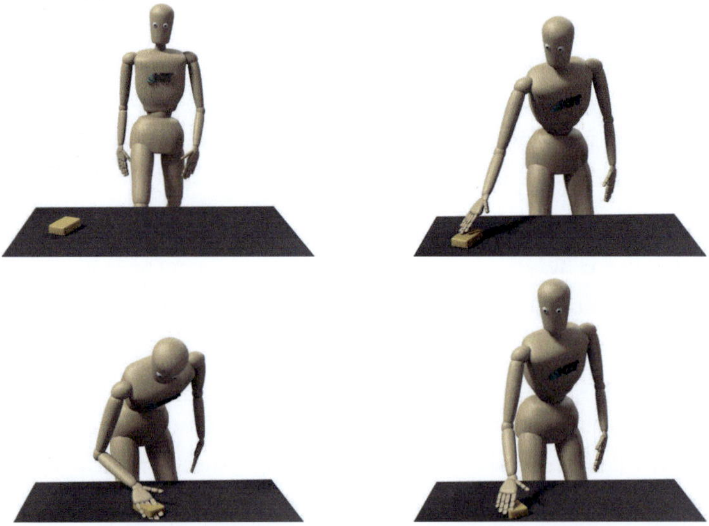

Figure 7.24: Snapshots of a demonstration of the wiping task.

7.4 End-to-End Use Case

The developed approach for end-to-end learning tasks from human observation and their execution on a humanoid robot is evaluated in the context of a table cleaning task. For the sake of clarity, a rather simple example is selected without limiting the generality of the approach. In the following, all steps involved in task learning and executing are described.

Observation

The first step is to observe the demonstration and map it to a representation suitable for learning. In the demonstrated task, the human wipes a table with a sponge. In the beginning, the sponge is placed on the table and the human is standing in front of the table. The human grasps the sponge with the right hand and a wiping action is started and executed for 13 s. After that, the human releases the sponge and retreats the hand. Figure 7.24 shows snapshots of the demonstration converted into the MMM reference model.

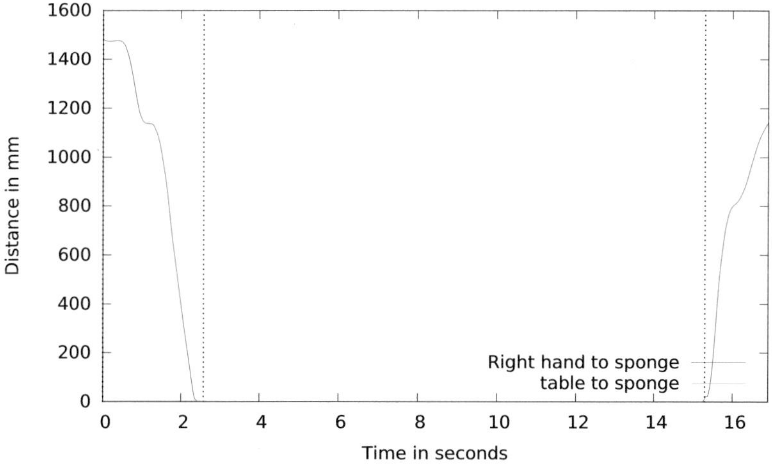

Figure 7.25: Semantic segmentation of a short wiping task. Semantic key frames (vertical dashed lines) are inserted when a stable contact is detected and when a contact is broken. The distance between the table and the sponge is almost zero during the whole demonstration.

Segmentation

The first processing step is the segmentation of the demonstration. The segmentation divides the demonstration into individual actions. Figure 7.25 shows the result of the semantic segmentation level of the hierarchical segmentation. The relevant distances between objects and the right hand are shown as curves. At the beginning, the distance between the right hand and the sponge is high (height $\approx$1.5 m), but decreases then rapidly to zero. It can be noticed that the first key frame is inserted slightly after the frame where the distance reaches zero since key frames are only inserted at a local minimum and if the relative velocity between two objects is almost constant. The first segment in Figure 7.25 corresponds to the grasp action, the middle segment corresponds to the wiping action and the last segment represents the retreat action.

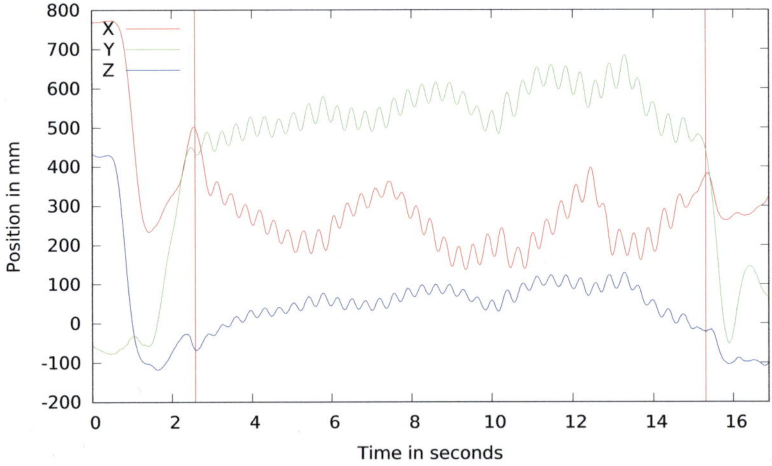

Figure 7.26: Hierarchical segmentation of a wiping task. The curves depict the motion of the right hand on the Cartesian axes whereas the red vertical dashed lines are semantic key frames.

Figure 7.26 shows the result of the complete hierarchical segmentation. The curves show the Cartesian positions of the right hand of the human. The motion characteristic based segmentation does not insert an additional key frame in this demonstration since it correctly identified that the wiping style was not changed during the execution. The amplitudes as well as the frequency in all dimensions of the motion are similar during the whole segment.

Recognition

The next step consists of the recognition of the extracted segments in the wiping task. An evaluation of the action recognition system developed in this thesis was described in subsection 7.1.9. In the wiping task, the semantic states at the beginning and end of each segment are calculated and converted into a matrix representation as described in section 4.5. These semantic states are shown in Figure 7.27. All object relations, in this demonstration only one, are also accumulated in the row and column labeled *object* to allow generalization. Before the *grasp* action (Figure 7.27(a)), only the sponge and the table are in contact. After the *grasp* action (Figure 7.27(b)), the hand is in contact with the sponge and the table as well. The contact with the sponge is desired, but the contact between the table and the hand results from the small height of the sponge and is not desired. After the *wiping* action (Figure 7.27(c)), the contact between the hand and the table is broken due to a different hand pose, but the contact between sponge and hand is still active. This relation illustrates, in this use case, the robustness of the action recognition and the task extraction to wrong results of previous processing steps. In addition to the semantic states, motion features are extracted from the segments and used for the recognition. The decision tree of the action recognizer was trained in advance with the whole dataset used in subsection 7.1.9. Based on these features, the decision tree of the action recognizer calculates the correct action labels: grasp, wiping and retreat.

Task Extraction

A new task can be extracted from the demonstration based on the recognized action labels, the semantic states from the segmentation and a database of known actions. Listing 7.1 shows all available actions for this use case: `grasp`, `lift`, `wiping`, `shaking` and `retreat`. It needs to be noted that not all predicates can be extracted from the semantic states of the segmentation. The predicates `cleaned` and `shaken` are not observable and can only be inferred based on executed actions. This needs to be considered when learning a new task with its preconditions and effects on the world state.

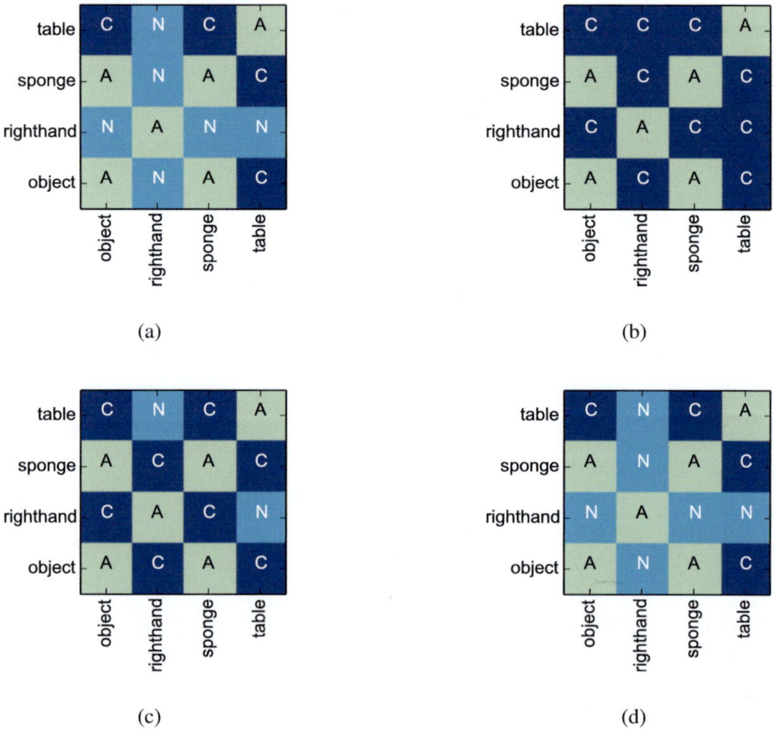

Figure 7.27: World states at key frames used for action recognition: (a) Before grasping the sponge (b) After grasping the sponge (c) End of wiping action (d) After retreating with the hand from the sponge

```
action grasp(h:hand, o:object, l:location):
  preconds:
    objectAt(o,l),
    handEmpty(h)
  effects:
    del(handEmpty(h)),
    add(inHand(o,h))

action lift(h:hand, o:object, l:location):
  preconds:
    objectAt(o,l),
    inHand(o,h)
  effects:
    del(objectAt(o,l))

action wiping(h:hand, s:sponge, l:location):
  preconds:
    objectAt(s,l),
    inHand(s,h)
  effects:
    add(cleaned(l))

action shaking(h:hand, b:bottle):
  preconds:
    inHand(b,h)
  effects:
    add(shaken(b))

action retreat(h:hand, o:object):
  preconds:
    objectAt(o,l),
    inHand(o,h)
  effects:
    del(inHand(o,h)),
    add(handEmpty(h))
```

Listing 7.1: Planning operators available for task learning.

The learning of a new task consisting of known actions is then solved by searching for action sequences with the same effects on the world state. This requires finding out which objects are involved in a task, which actions explain the observation best and how these actions need to be parametrized. The action recognition provides a sequence of labeled actions, but these labels are not correct if the action recognition failed. Therefore, the recognized actions are only considered as potential action candidates. The recognized actions are used as the preferred solution if it does not contradict any of the preconditions or effects. Otherwise, all available actions are validated with respect to the observed world state changes. The first matching action is selected for the new task. The learned task for this use case is shown in Listing 7.2.

```
action WipingTask(sponge:sponge,
    location:location, hand:hand):
preconds:
objectAt(sponge, location),
handEmpty(hand)
effects:
add(cleaned(location))
action list:
grasp(hand, sponge, location),
wiping(hand, sponge, location),
retreat(hand, sponge)
```

Listing 7.2: Extracted planning operator.

Based on the extracted action sequence it is also possible to extract the types for the parameters from the action. In this use case, the first parameter needs to be an object of type `sponge` since the wiping action requires a sponge. The second parameter is the `location` that should be wiped and the third parameter can be the left or right `hand`. The extracted preconditions of the task are that the `sponge` is at `location` and that the `hand` to be used is empty. The effect of the task is that the location is `cleaned`. This new task is then added to the OAC database in the long-term memory and can be used by the planner in the future.

Robot Skills

Each of the planning operators needs an implementation for execution by a robot. These planning operators are implemented using the statechart concept presented in this thesis. The statecharts of the actions required for the execution of the new task are shown in Figure 7.28. The *grasp* statechart uses visual servoing to move the hand of the robot to a grasping pose relative to the current object pose. The grasping information is stored for each object in the prior knowledge. Upon reaching the grasping pose, the hand is closed. The *wiping* statecharts uses a Dynamic Movement Primitive learned from observation (Gams et al., 2010) and the *retreat* action opens the hand and moves it upwards and towards the robot.

Execution based on goal query

As mentioned before, the newly learned task can be used by a symbolic planner like any other planning operator. This leads to shorter plans as the learned planning operator already consists of a sequence of elementary actions. This reduces the computational complexity of planning since this complexity grows exponentially with the plan length and thus results in faster planning.

To trigger the use of the new task, the following task goal could be used:

`forallK(?l : location) K(cleaned(?l))`

This goal state specifies that the predicate `cleaned` should be true for all constants of type `location`. Figure 7.29 shows snapshots of the execution of the new task in simulation, in which the robot is standing in front of the table and the sponge is on the table.

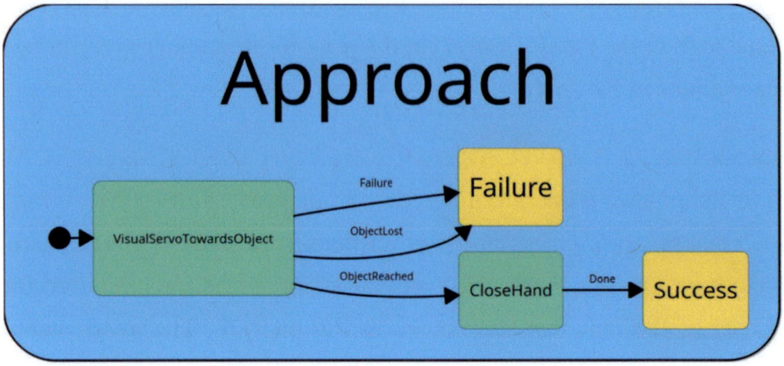

(a)

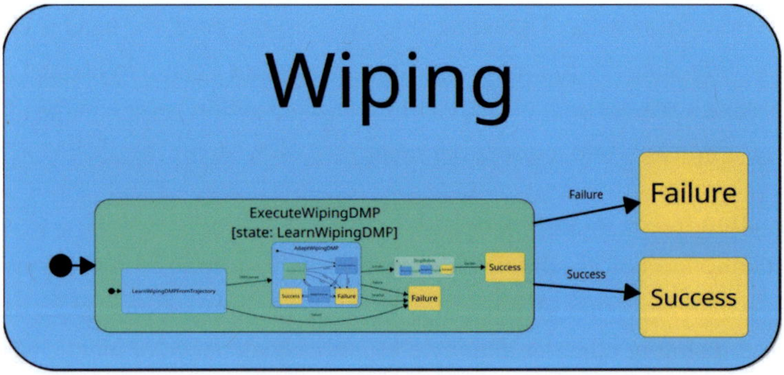

(b)

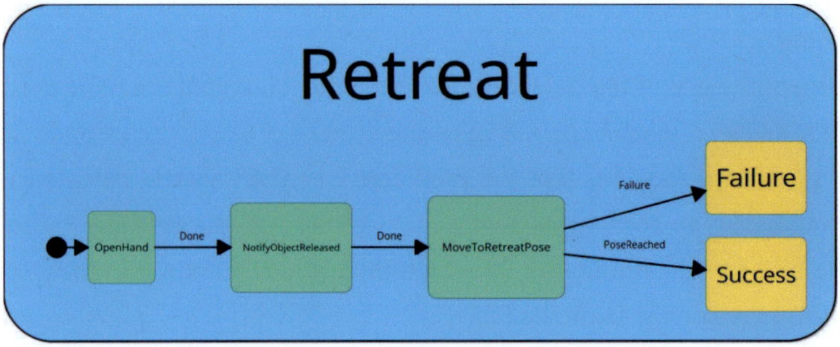

(c)

Figure 7.28: Robot skills required for the wiping task modeled as statecharts.

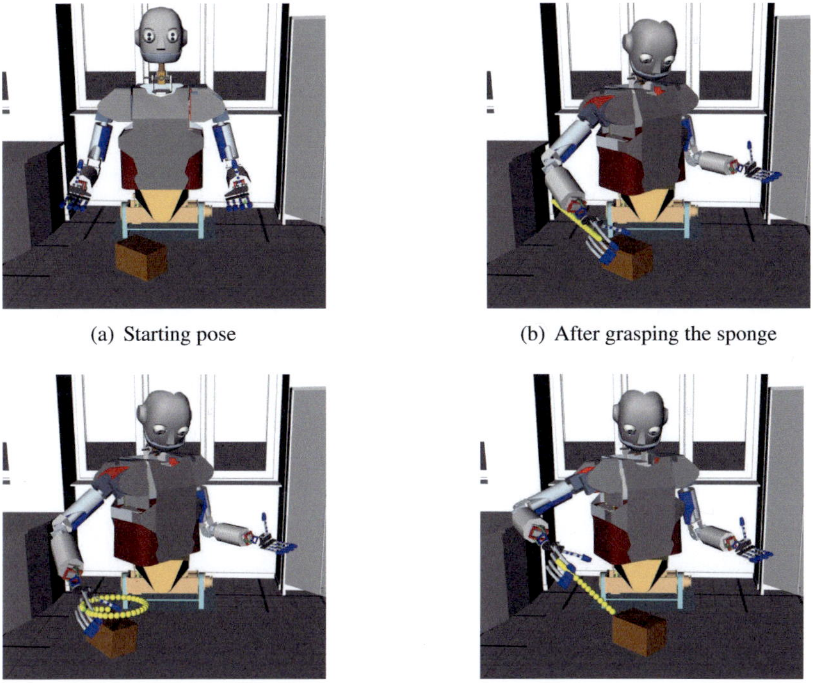

(a) Starting pose

(b) After grasping the sponge

(c) Execution of the wiping action

(d) After retreating the hand from the sponge

Figure 7.29: Snapshots of the execution of the new wiping task. The yellow lines depict the trajectory of the end-effector.

7.5 Summary

In this chapter, the presented approaches regarding task understanding by segmentation and action recognition, robot skill modeling with statecharts and task solving and execution were evaluated. The hierarchical segmentation algorithm was evaluated on a demonstration dataset containing typical household manipulation actions such as pick, place, pour, stir and cut. Its performance was compared to two popular approaches and to a manually created reference segmentation. The results and a user study demonstrate that the automatically extracted segments by the developed algorithms are similar to the way humans segment a demonstration. The action recognition approach was trained on one part of the same dataset and evaluated on another part with cross-validation. The results show that the semantic and the motion feature space complement each other and allow the classification of actions that are ambiguous in one of the feature spaces.

The robot skill modeling approach based on statecharts was employed and evaluated by letting multiple developers implement a multitude of skills using different sensor modalities and targeting different abstraction levels such as object grasping or robot navigation. Developer feedback showed that the approach simplified the design process significantly and improved the maintainability.

The task solving and execution approach was applied in a kitchen scenario, in which several tasks have to be executed such as setting a table or preparing dinner for two people by utilizing the developed robot skills. Furthermore, a user study was conducted to evaluate if untrained users are able to instruct the robot with natural language to solve given tasks. The results indicate that most users are able to solve the tasks, but multiple trials are necessary.

8 Conclusion

The objective of this thesis was to develop methods and algorithms for learning complex manipulation tasks from human demonstration and execute them on a humanoid robot in dynamic environments. The proposed approach addresses the three major research questions in learning from human observation and robot programming by demonstration: 1) Observation and understanding of human demonstrations, 2) representation of skills and tasks in a generalized way and 3) adaptation and execution of learned skill and task knowledge in novel situations. Human demonstrations of complex manipulation tasks are automatically segmented into meaningful segments and associated with known actions. These actions are modeled with statecharts for execution on mobile, dual-arm robots such as the humanoid robot ARMAR-III. Finally, the actions are used to solve tasks in dynamic environments including reasoning about potential object alternatives and monitoring of the execution. In this chapter, the contributions of this work are revisited and the results as well as possible extensions are discussed.

8.1 Contributions of the Thesis

The overarching contribution of the thesis is a complete system for learning task knowledge from human demonstrations and the adaptation of this knowledge to new situations. In summary, the main contributions of this thesis are:

Task Understanding by Segmentation and Action Recognition

To understand and learn complex manipulation tasks demonstrated by humans, it is necessary to segment these demonstrations. To this end, a novel hierarchical segmentation approach is proposed to extract meaningful segments, i.e. manipulation actions, that takes not only motion data, but also semantic information extracted from object contact relations into consideration. For the recognition of such manipulation actions in human demonstrations, an action recognizer has been developed which is based on a decision tree classifier. This classifier relies on a novel action descriptor considering object relations and global motion features. The hierarchical segmentation approach has been published first in (Wächter et al., 2013) and was extended in (Wächter and Asfour, 2015). It is used in (Wörgötter et al., 2015) and (Schlichting et al., 2018).

Robot Skill Modeling with Statecharts

A novel robot skill modeling approach using extended hierarchical statecharts has been developed, implemented and evaluated in complex tasks on a humanoid robot. This hierarchical modeling approach is capable of representing multi-modal robot skills on all abstraction levels of a robot architecture ranging from low-level controllers to high-level symbolic planning operators. The developed hierarchical statechart approach extends the original statecharts described in (Harel, 1987) to address the requirements of many-sensor and many-actor systems such as humanoid robots by providing means for fine-granular and reusability-supporting data flow specification, distribution over multiple hosts and fault handling on coordination level. On top of this, a complete graphical modeling tool has been developed to support convenient programming, easy system monitoring, simplified debugging and short development cycles in robot programming. The statechart concept and the graphical tool are core elements of the robot development environment

ArmarX[1], which is used for programming the ARMAR robots. The concept and its integration in ArmarX has been published in (Welke et al., 2013c; Vahrenkamp et al., 2014, 2015) and (Wächter et al., 2016).

Task Solving and Execution

For the execution of tasks in dynamic environments symbolic planning is used to find an action sequence that transforms an observed world state into the goal state of a task. For this purpose, continuous, subsymbolic sensorimotor representations are generically mapped into a symbolic representation of the world state. To deal with situations in which the world state is insufficiently explored or objects are unknown, a symbol replacement method has been proposed to generate hypotheses about alternatives for objects and object locations to allow the generation of a plan for solving the underlying task. In order to handle execution failures and wrong hypotheses, the execution of tasks is monitored and corrected automatically. The task solving and execution approach has been published in (Ovchinnikova et al., 2015). An extended version with replacement capabilities is presented in (Wächter et al., 2018).

8.2 Discussion and Outlook

Several aspects of learning from demonstration have been addressed in this thesis. In the following, the results of the proposed approach and possible extensions are discussed.

Task Segmentation and Action Recognition

The evaluation of the hierarchical segmentation approach showed that the extracted segments are in many cases similar to segments defined by humans. However, there are cases where segmentation differs due to various reasons.

[1] ArmarX: `https://armarx.humanoids.kit.edu`

Semantic segmentation

In general, the semantic segmentation approach heavily relies on the precision of the 6D trajectory and the models associated with it. An inaccurate model can lead to wrong or missed contact detection and therefore to wrong segmentation. Based on experience, the precision is sufficient to detect all contacts and segment the trajectory into its semantic parts. Although the semantic segmentation performs well for most demonstrations, the drawback of the semantic segmentation becomes apparent in the case of actions without observable effects. In this case, the semantic segmentation has little or no chance to detect these actions.

Segmentation based on Motion Characteristic

The sub-segmentation tackles this problem of unobservable effects of actions. Additionally, different styles of periodic actions can be detected such as different wiping styles, e.g. wiping in lines or intensively wiping on one spot. In the experiments, most segments have been detected, though a smooth transition between two actions can be problematic and no key frame might be found.

Hierarchical Segmentation

The benefit of combining both segmentation algorithms to a hierarchical segmentation was demonstrated by using specialized datasets. The semantic segmentation performs naturally well on demonstrations with observable effects whereas the motion characteristic based segmentation performs better on demonstrations containing periodic motions. The hierarchical segmentation outperforms both algorithms on both used datasets. Both algorithms have their strengths and weaknesses, which are compensated by their combination. The semantic segmentation, and thus the hierarchical segmentation, also provides valuable information about the semantic state of each segment, which a motion based segmentation cannot provide. While this information is not important for the segmentation results themselves, it is particularly valuable for further processing of the segmentation results such as action recognition or learning of effects of actions.

Action Recognition

In contrast to the state of the art, the proposed approach uses motion and semantic information to recognize actions. The combination of both feature spaces has proven to be advantageous as actions with very similar motions can be reliably distinguished (*tossing* vs. *wiping*), which are difficult to correctly recognize by using classical action recognition based on motion features. Yet motions without semantic change and no characteristic motion are not reliably recognized.

The decision tree classifier is able to recognize irrelevant variances in the demonstrations. For example, in some demonstrations the hand touches the table at the end of the *approach* action. This contact is not relevant for this action, which is successfully recognized by the classifier.

As for all data-driven approaches, more training data would have helped the classifier to create a better decision tree since many different actions are to be recognized and since each action type increases the ambiguity. In future work, more object relations such as *in* and *on* could improve the action recognition results further since they reduce the ambiguity of actions.

Task Extraction

The successful extraction of new tasks learned from human observation depends on the correctness of the segmentation and recognition. One missed or additional segment will cause a shorter or longer action sequence for the new task. Furthermore, the action recognition has to recognize an action that should not be contained in the extracted action sequence. Depending on the recognized action, this can lead to a failure of the task extraction algorithm or even lead to a task description with different semantics. In future work, this problem could be addressed by learning task representations from multiple demonstrations. Another future extension could be pruning of generated plans. Using planning operators learned from demonstration creates the possibility that the resulting plan is not optimal. For example, a learned task contains *grasp* as a first action, but the robot already holds the object in its hand. The planner will then instruct a *place* action to *grasp* it again as part of the learned planning operator. Such cases could be detected and removed from the action sequence by detecting which consecutive actions nullify the effects of the previous action.

Robot Skill Modeling with Statecharts

In the following, experiences from implementing and developing robot skills with the statechart concept are discussed. Since more than 300 states for a wide variety of applications for the robots of the ARMAR series are already developed, several advantages and disadvantages of the proposed concept appeared. The presented statechart approach has extensively been used not only to program simple actions, but also for complex skills applied in real world scenarios, including grasping, mixing, pouring or opening and closing doors.

The decision to remove some features of Harel's original statechart has proved to be the right decision, since the removed features (inter-level-transitions, history-connector) were rarely missed in robotic applications and their removal has led to significant improvements regarding comprehension, reusability and maintenance. A detailed and explicit specification of the data flow results in development overhead, but the simplified debugging and maintenance makes the effort worth in the long run. Specifying and inspecting this data flow with graphical tools greatly simplifies the software development process. Such graphical tools are not only useful for defining the data flow, but indispensable for developing complex state machines. Hence, the graphical statechart editor is one of the most important tools of the ArmarX framework.

The proposed statechart concept proved to be applicable for different abstraction levels from low-level controllers to high-level planning operators as well as for heterogeneous skills such as grasping and walking. In future work, extending the proposed statechart concept with full orthogonality with consideration of splitting and joining the data flow could simplify developing robot capabilities, in which skills and tasks are executed in parallel as it is the case in bimanual grasping and manipulation.

Task Solving and Execution

With the presented task solving and execution approach it is possible to equip a robot with the powerful capability to act and interact autonomously in a dynamic environment. The ability to reason about probable locations of objects or entire replacements of objects greatly enhances the autonomy and flexibility of a robot. Yet the feasibility of a solution heavily depends on the reliability of the used information. Up to now, the information about the certainty of existence of objects is not incorporated into the planning domain generation. Thus, every object pose is effectively treated as a fact. However, the full usage of such information would require a probabilistic planner.

In future work, the conversion of continuous sensorimotor information into symbolic information and vice versa could be extended to be refined or learned through experience and exploration. Refinement could be achieved through evaluation of previous action results, where failure and success of actions can be used to tune hyper parameters and confidence values.

The user study showed that the proposed approach can be controlled by untrained users to solve a complex task with a humanoid robot. The problems that the participants encountered showed that the feedback of the robot, i.e. the human-robot-interaction, should be improved in future extensions.

A Planning Domain

This section shows a shortened planning domain in the PKS (*Planning with Knowledge and Sensing*, (Petrick and Bacchus, 2002)) syntax as generated by the implementation of the task solving and execution approach presented in chapter 6. The domain is shown in listing A.1.

A domain consists of three parts: the symbols definition, a list of actions and a problem definition. The symbols definition consists itself as well of 3 parts: the `types` declaration, the `predicates` declaration and the `constants` declaration. The declaration of the used types, e.g. `agent_robot`, determines the types that are used for action parameters and constants. In the generated example domain each type is prefixed with a meta type, e.g. `location_*`, which is necessary for the conversion back into robot memory symbols.

The `predicates` declaration determines the predicates that can be used in the domain and have the arity attached, e.g. `agentAt/2` means that the predicate always has two parameters, for example `agentAt(robot_IE83, fridge_DE32)`.

The `constants` declaration represents the entities existing in the planning domain together with the type of the constant.

The `actions` consist of a typed parameter list, preconditions and effects. Preconditions are predicates that can only contain the parameters of the action and effects are predicate changes. The `problem` definition depends on the current state of the world and describes the state of the world with grounded predicates in the `initdb` block. The second part of the `problem` definition is the `goal` definition. The `goal` definition is a desired state of an arbitrary number of predicates. To be more flexible quantifiers can be used, which is particular important for generated domains. The goal in the example in listing A.1 specifies that all doors should be closed, i.e. *not* `open`. Without quantifiers each `location` constant would have to be named explicitly. If a generated domain with varying locations is used, this would entail that the `goal` would to generated dynamically as well. This can be avoided with quantifiers as shown in the example.

```
domain reduced-kitchen-setting {

  # available symbols

  symbols {
    types:
    agent_robot,
    location,
    location_doors,
    obj_graspable,
    obj_hand,
    obj_corn;

    predicates:
    agentAt/2,
    objectAt/2,
    handEmpty/2,
    open/1;

    constants:
    agent_robot robot_IE83;
    location_doors fridge_DE32;
    obj_graspable corn_FE15;
    obj_graspable juice_JK98;
    obj_hand lefthand_FE99, righthand_FG65;
  }


  # action definitions
```

```
action close(?a : agent_robot, ?d :
    location_doors, ?h : obj_hand) {
  preconds:
    K(open(?d)) &
    K(agentAt(?a, ?d)) &
    K(handEmpty(?a, ?h))
  effects:
    del(Kf, open(?d))
}

action move(?a : agent_robot, ?l1 :
    location, ?l2 : location) {
  preconds:
    K(agentAt(?a, ?l1)) &
    K(?l1 != ?l2) &
  effects:
    add(Kf, agentAt(?a, ?l2)),
    del(Kf, agentAt(?a, ?l1))
}


# problem definitions
# (initial state & goal)

problem default-problem {
  initdb {
    Kf:
    agentAt(robot_IE83, fridge_DE32),
    handEmpty(robot_IE83, lefthand_FE99),
    handEmpty(robot_IE83, righthand_FG65),
    objectAt(juice_JK98, fridge_DE32)
    objectAt(corn_FE15, fridge_DE32),
    open(fridge_DE32)

  }

  goal:
```

```
            (forallK(?l : location_doors)
                !K(open(?l)))
        }
    }
```

Listing A.1: An exemplary definition of a planning domain.

Glossary

action An action is a targeted motion performed by one subject, e.g. a step during walking or grasping an object.

affordance An affordance (Gibson, 1979) is the possibility of an action on an object or environment based on various properties such as shape, weight, stability etc. For example, a chair affords *sitting*, but it does not afford *rolling*.

causal A causal algorithm only uses data from the current state and the past. It does not need data from future timestamps. This is a crucial criterion for an online algorithm.

joint space The joint space of a robot is the configuration space, where the joint angle of each joint represent the dimensions of the configuration space.

key frame A motion consists of a sequence of frames. A key frame in the context of segmentation is a frame that denotes the end and beginning of two consecutive segments.

motion primitive A motion primitive is the smallest unit of motions with a semantic meaning, e.g. a step while walking could be one motion primitive.

Object-Action Complex Object-Action Complex (OAC) is a manipulation action formalization to describe manipulation actions in relation to an object and the effect the action has on the world state.

observer An ArmarX observer is a component that observes and logs the data stream of one sensor type. It is possible to install conditions and filters on each data field of an observer that are evaluated on each sensor update.

planning operator A planning operator is the equivalent of an action in terms of the planning system, i.e. it contains the information needed for planning: the precondition predicates, the effects and the types of all involved planning constants.

predicate Predicates are used to symbolically describe the state of the world. They have a Boolean value and zero to n parameters, e.g. `on(cup, table)`.

robot skill A robot skill is the implementation of an action on a robot, e.g. *moving to a landmark* or *grasping an object*.

segment A (motion) segment is a part of a longer demonstration which ideally denotes one motion primitive.

semantic segment A semantic segment is a segment of a demonstration that contains a semantic change in the world, e.g. the change of contact between two objects.

task A task is a goal for the robot that should be achieved by executing the correct actions. The goal is formulated as a desired goal world state.

task space The task space of a robot is the configuration space of the 6D end-effector pose of a robot arm.

world state The world state means in this thesis a semantic or symbolic representation of the environment. More specifically, in the case of the semantic segmentation it refers to object relations and in case of the symbolic planning to a set of predicates describing the relations between entities such as objects and robots.

Acronyms

DMP Dynamic Movement Primitives

DNA Deoxyribonucleic acid

DOF Degree of Freedom

FSM Finite-state Machine

GUI Graphical User Interface

HMM Hidden Markov Model

HS Hierarchical Segmentation

HTN Hierarchical Task Network

IDE Integrated Development Environment

IDL Interface Definition Language

IMU Inertial Measurement Unit

KIT Karlsruhe Institute of Technology

LfD Learning from Demonstration

MMM Master Motor Map

OAC	Object-Action Complex
PbD	Programming by Demonstration
PCA	Principal Component Analysis
PD	Proportional Derivative (-Controller)
PKS	Planning with Knowledge and Sensing
rFSM	restricted Finite State Machines
RGB	Red Green Blue
RGB-D	Red Green Blue Depth
ROS	Robot Operating System
SEC	Semantic Event Chains
SRM	Symbol Replacement Manager
UML	Unified Modeling Language
XML	Extensible Markup Language
ZVC	Zero Velocity Crossings

List of Figures

List of Tables

Bibliography

Aein, M. J., Aksoy, E. E., Tamosiunaite, M., Papon, J., Ude, A., and Wörgötter, F. (2013). Toward a library of manipulation actions based on semantic object-action relations. In *2013 IEEE/RSJ International Conference on Intelligent Robots and Systems*, pages 4555–4562. IEEE.

Aein, M. J., Aksoy, E. E., and Wörgötter, F. (2017). Library of Actions: Implementing a Generic Robot Execution Framework by Using Manipulation Action Semantics. *The International Journal of Robotics Research (submitted)*.

Aggarwal, J. K. and Ryoo, M. S. (2011). Human activity analysis: A review. *ACM Computing Surveys (CSUR)*, 43(3):16:1–16:43.

Agostini, A., Aein, M. J., Szedmak, S., Aksoy, E. E., Piater, J., and Wörgötter, F. (2015). Using structural bootstrapping for object substitution in robotic executions of human-like manipulation tasks. In *Intelligent Robots and Systems (IROS), 2015 IEEE/RSJ International Conference On*, pages 6479–6486. IEEE.

AIST (2017). RtcLink [software].

Aksoy, E. E., Abramov, A., Dörr, J., Ning, K., Dellen, B., and Wörgötter, F. (2011). Learning the semantics of object–action relations by observation. *The International Journal of Robotics Research*, 30(10):1229–1249.

Aksoy, E. E., Abramov, A., Wörgötter, F., and Dellen, B. (2010). Categorizing object-action relations from semantic scene graphs. In *Robotics and Automation (ICRA), 2010 IEEE International Conference On*, pages 398–405. IEEE.

Aksoy, E. E., Orhan, A., and Wörgötter, F. (2016). Semantic Decomposition and Recognition of Long and Complex Manipulation Action Sequences. *International Journal of Computer Vision*.

Aksoy, E. E., Ovchinnikova, E., Orhan, A., Yang, Y., and Asfour, T. (2017). Unsupervised linking of visual features to textual descriptions in long manipulation activities. *IEEE Robotics and Automation Letters (RA-L) with ICRA presentation*, 2(3):1397–1404.

Aksoy, E. E., Tamosiunaite, M., and Wörgötter, F. (2015). Model-free incremental learning of the semantics of manipulation actions. *Robotics and Autonomous Systems*, 71:118–133.

Amft, O., Junker, H., and Tröster, G. (2005). Detection of eating and drinking arm gestures using inertial body-worn sensors. *Proc. 9th IEEE Symp. Wearable Comput.*, pages 160–163.

Angermann, A., Beuschel, M., Rau, M., and Wohlfarth, U. (2014). *MATLAB-Simulink-Stateflow: Grundlagen, Toolboxen, Beispiele*. Walter de Gruyter.

Aoki, T., Venture, G., and Kulić, D. (2013). Segmentation of human body movement using inertial measurement unit. In *Proceedings - 2013 IEEE International Conference on Systems, Man, and Cybernetics, SMC 2013*, pages 1181–1186.

Argall, B., Chernova, S., Veloso, M., and Browning, B. (2009). A survey of robot learning from demonstration. *Robotics and Autonomous Systems*, 57(5):469–483.

Arkin, R. C. (1998). *Behavior-Based Robotics*. MIT press.

Asfour, T. (2003). *Sensomotorische Bewegungskoordination zur Handlungsausführung eines humanoiden Roboters*. PhD thesis, Universität Karlsruhe, Karlsruhe.

Asfour, T., Azad, P., Gyarfas, F., and Dillmann, R. (2008). Imitation Learning of Dual-Arm Manipulation Tasks in Humanoid Robots. *International Journal of Humanoid Robotics*, 5(2):183–202.

Asfour, T., Regenstein, K., Azad, P., Schröder, J., N. Vahrenkamp, and Dillmann, R. (2006). ARMAR-III: An Integrated Humanoid Platform for Sensory-Motor Control. In *IEEE/RAS International Conference on Humanoid Robots (Humanoids)*, pages 169–175, Genova, Italy.

Asfour, T., Schill, J., Peters, H., Klas, C., Bücker, J., Sander, C., Schulz, S., Kargov, A., Werner, T., and Bartenbach, V. (2013). ARMAR-4: A 63 DOF Torque Controlled Humanoid Robot. In *IEEE/RAS International Conference on Humanoid Robots (Humanoids)*, pages 390–396.

Ataya, A., Jallon, P., Bianchi, P., and Doron, M. (2013). Improving activity recognition using temporal coherence. *Engineering in Medicine and Biology Society (EMBC), 2013 35th Annual International Conference of the IEEE*, 2013:4215–4218.

Atkinson, R. C. and Shiffrin, R. M. (1968). Human memory: A proposed system and its control processes. *Psychology of learning and motivation*, 2:89–195.

Azad, P., Asfour, T., and Dillmann, R. (2008). Robust Real-time Stereo-based Markerless Human Motion Capture. In *IEEE/RAS International Conference on Humanoid Robots (Humanoids)*, pages 700–707.

Bandouch, J. and Beetz, M. (2009). Tracking humans interacting with the environment using efficient hierarchical sampling and layered observation models. In *Computer Vision Workshops (ICCV Workshops), 2009 IEEE 12th International Conference On*, pages 2040–2047. IEEE.

Barbic, J., Safonova, A., Pan, J. Y., Faloutsos, C., Hodgins, J. K., and Pollard, N. S. (2004). Segmenting motion capture data into distinct behaviors. In *GI '04: Proceedings of Graphics Interface 2004*, pages 185–194, School of Computer Science, University of Waterloo, Waterloo, Ontario, Canada. Canadian Human-Computer Communications Society.

Barchunova, A., Haschke, R., Grossekathoefer, U., Wachsmuth, S., Janssen, H., and Ritter, H. (2011). Unsupervised Identification of Object Manipulation Operations from Multimodal Input. In *Workshop New Challenges in Neural Computation 2011*, page 42. Citeseer.

Bashir, F., Qu, W., Khokhar, A., and Schonfeld, D. (2005). HMM-based motion recognition system using segmented PCA. *Proc. IEEE Conf. Image Process.*, pages 1288–1291.

Beaudoin, P., Coros, S., van de Panne, M., and Poulin, P. (2008). Motion-motif graphs. In *Proceedings of the 2008 ACM SIGGRAPH/Eurographics Symposium on Computer Animation*, pages 117–126. Eurographics Association.

Beetz, M., Klank, U., Kresse, I., Maldonado, A., Mosenlechner, L., Pangercic, D., Ruhr, T., and Tenorth, M. (2011). Robotic roommates making pancakes. In *Proc. of Humanoids*, pages 529–536.

Beetz, M., Tenorth, M., and Winkler, J. (2015). Open-EASE. In *Robotics and Automation (ICRA), 2015 IEEE International Conference On*, pages 1983–1990. IEEE.

Bentley, J. L. (1975). Multidimensional Binary Search Trees Used for Associative Searching. *Commun. ACM*, 18(9):509–517.

Berlin, E. and Van Laerhoven, K. (2012). Detecting leisure activities with dense motif discovery. In *UbiComp'12 - Proceedings of the 2012 ACM Conference on Ubiquitous Computing*, pages 250–259.

Besl, P. J. and McKay, N. D. (1992). A Method for Registration of 3-D Shapes. *IEEE Trans. Pattern Anal. Mach. Intell.*, 14(2):239–256.

Billington, D., Estivill-Castro, V., Hexel, R., and Rock, A. (2010). Modelling Behaviour Requirements for Automatic Interpretation, Simulation and Deployment. In *SIMPAR*, volume 6472 of *Lecture Notes in Computer Science*, pages 204–216. Springer.

Bohren, J., Rusu, R. B., Jones, E. G., Marder-Eppstein, E., Pantofaru, C., Wise, M., Mösenlechner, L., Meeussen, W., and Holzer, S. (2011). Towards autonomous robotic butlers: Lessons learned with the pr2. In *Robotics and Automation (ICRA), 2011 IEEE International Conference On*, pages 5568–5575. IEEE.

Bollini, M., Tellex, S., Thompson, T., Roy, N., and Rus, D. (2013). Interpreting and executing recipes with a cooking robot. In *Experimental Robotics*, pages 481–495. Springer.

Boteanu, A., St. Clair, A., Mohseni-Kabir, A., Saldanha, C., and Chernova, S. (2016). Leveraging Large-Scale Semantic Networks for Adaptive Robot Task Learning and Execution. *Big Data*, 4(4):217–235.

Bradski, G. R. and Davis, J. W. (2002). Motion segmentation and pose recognition with motion history gradients. *Machine Vision and Applications*, 13(3):174–184.

Breen, M. (2004). Statecharts: Some Critical Observations.

Breiman, L. (1996). Bagging predictors. *Machine learning*, 24(2):123–140.

Bruyninckx, H., Soetens, P., and Koninckx, B. (2003). The Real-Time Motion Control Core of the Orocos Project. In *IEEE International Conference on Robotics and Automation (ICRA)*, pages 2766–2771.

Burges, C. J. (1998). A tutorial on support vector machines for pattern recognition. *Data mining and knowledge discovery*, 2(2):121–167.

Burns, A., Greene, B. R., McGrath, M. J., O'Shea, T. J., Kuris, B., Ayer, S. M., Stroiescu, F., and Cionca, V. (2010). SHIMMER™–A wireless sensor platform for noninvasive biomedical research. *IEEE Sensors Journal*, 10(9):1527–1534.

Chamroukhi, F., Mohammed, S., Trabelsi, D., Oukhellou, L., and Amirat, Y. (2013). Joint segmentation of multivariate time series with hidden process regression for human activity recognition. *Neurocomputing*, 120:633–644.

Chang, G. and Kulić, D. (2013a). Robot task error recovery using Petri nets learned from demonstration. In *Advanced Robotics (ICAR), 2013 16th International Conference On*, pages 1–6. IEEE.

Chang, G. and Kulić, D. (2013b). Robot task learning from demonstration using petri nets. In *2013 IEEE RO-MAN*, pages 31–36. IEEE.

Coleman, D., Hayes, F., and Bear, S. (1992). Introducing objectcharts or how to use statecharts in object-oriented design. *Software Engineering, IEEE Transactions on*, 18(1):8–18.

Devanne, M., Berretti, S., Pala, P., Wannous, H., Daoudi, M., and Del Bimbo, A. (2017). Motion segment decomposition of RGB-D sequences for human behavior understanding. *Pattern Recognition*, 61:222–233.

Dillmann, R. (2004). Teaching and Learning of Robot Tasks via Observation of Human Performance. *Robotics and Autonomous Systems*, 47(2-3):109–116.

Dzifcak, J., Scheutz, M., Baral, C., and Schermerhorn, P. (2009). What to do and how to do it: Translating natural language directives into temporal and dynamic logic representation for goal management and action execution. In *Proc. of ICRA*, pages 4163–4168.

EasyCODE (2015). EasyCODE. `http://www.easycode.de`.

Ekvall, S. and Kragic, D. (2006). Learning task models from multiple human demonstrations. In *ROMAN 2006-The 15th IEEE International Symposium on Robot and Human Interactive Communication*, pages 358–363. IEEE.

Ekvall, S. and Kragic, D. (2008). Robot learning from demonstration: A task-level planning approach. *International Journal of Advanced Robotic Systems*, 5(3):223–234.

Erol, K., Hendler, J., and Nau, D. S. (1994). HTN planning: Complexity and expressivity. In *AAAI*, volume 94, pages 1123–1128.

Fearnhead, P. (2006). Exact and efficient Bayesian inference for multiple changepoint problems. *Statistics and computing*, 16(2):203–213.

Fikes, R. E. and Nilsson, N. J. (1971). STRIPS: A new approach to the application of theorem proving to problem solving. *Artificial intelligence*, 2(3-4):189–208.

Fod, A., Matarić, M. J., and Jenkins, O. C. (2002). Automated derivation of primitives for movement classification. *Autonomous robots*, 12(1):39–54.

Fox, E., Hughes, M., Sudderth, E., and Jordan, M. (2014). Joint modeling of multiple time series via the beta process with application to motion capture segmentation. *Annals of Applied Statistics*, 8(3).

Frank, M., Leitner, J., Stollenga, M., Harding, S., Förster, A., and Schmidhuber, J. (2012). The Modular Behavioral Environment for Humanoids and other Robots (MoBeE). In *ICINCO (2)*, pages 304–313. Citeseer.

Fraser, L., Rekabdar, B., Nicolescu, M., Nicolescu, M., Feil-Seifer, D., and Bebis, G. (2016). A compact task representation for hierarchical robot control. In *2016 IEEE-RAS 16th International Conference on Humanoid Robots (Humanoids)*, pages 697–704.

Freund, Y., Schapire, R., and Abe, N. (1999). A short introduction to boosting. *Journal-Japanese Society For Artificial Intelligence*, 14(771-780):1612.

Gams, A., Do, M., Ude, A., Asfour, T., and Dillmann, R. (2010). On-line periodic movement and force-profile learning for adaptation to new surfaces. In *Humanoid Robots (Humanoids), 2010 10th IEEE-RAS International Conference On*, pages 560–565.

Geib, C., Mourao, K., Petrick, R., Pugeault, N., Steedman, M., Krueger, N., and Wörgötter, F. (2006). Object action complexes as an interface for planning and robot control. In *IEEE RAS International Conference on Humanoid Robots*.

Geib, C. W. (2009). Delaying Commitment in Plan Recognition Using Combinatory Categorial Grammars. In *Proc. of IJCAI*, pages 1702–1707.

Gibson, J. J. (1979). *The Ecological Approach to Visual Perception: Classic Edition*. Houghton, Mifflin and Company.

Gill, A. (1962). *Introduction to the Theory of Finite-State Machines*. McGraw-Hill, New York, NY, USA.

Guerra-Filho, G. and Aloimonos, Y. (2007). A language for human action. *Computer*, 40(5):42–51.

Han, J., Shao, L., Xu, D., and Shotton, J. (2013). Enhanced computer vision with microsoft kinect sensor: A review. *IEEE transactions on cybernetics*, 43(5):1318–1334.

Hao, Y., Chen, Y., Zakaria, J., Hu, B., Rakthanmanon, T., and Keogh, E. (2013). Towards never-ending learning from time series streams. *Proc. ACM Conf. Knowl. Discovery Data Mining*, pages 874–882.

Harel, D. (1987). Statecharts: A visual formalism for complex systems. *Science of computer programming*, 8(3):231–274.

Harel, D. and Politi, M. (1998). *Modeling Reactive Systems with Statecharts: The STATEMATE Approach*. McGraw-Hill, Inc.

Hendrich, N., Klimentjew, D., and Zhang, J. (2010). Multi-sensor based segmentation of human manipulation tasks. In *Multisensor Fusion and Integration for Intelligent Systems (MFI), 2010 IEEE Conference On*, pages 223–229. IEEE.

Henning, M. (2004). A new approach to object-oriented middleware. *Internet Computing, IEEE*, 8(1):66–75.

Hirzinger, G. and Bauml, B. (2006). Agile robot development (aRD): A pragmatic approach to robotic software. In *Intelligent Robots and Systems, 2006 IEEE/RSJ International Conference On*, pages 3741–3748. IEEE.

Hoffmann, J. and Nebel, B. (2001). The FF planning system: Fast plan generation through heuristic search. *Journal of Artificial Intelligence Research*, 14:253–302.

Huber, A. (2007). Boost statechart library. `http://www.boost.org`.

Ilg, W., Bakir, G., Mezger, J., and Giese, M. (2004). On the representation, learning and transfer of spatio-temporal movement characteristics. *Int. J. Human Robot.*, 1:613–636.

Jain, A. K., Duin, R. P. W., and Mao, J. (2000). Statistical pattern recognition: A review. *IEEE Transactions on pattern analysis and machine intelligence*, 22(1):4–37.

Jain, A. K., Mao, J., and Mohiuddin, K. M. (1996). Artificial neural networks: A tutorial. *IEEE computer*, 29(3):31–44.

Jäkel, R., Meissner, P., Schmidt-Rohr, S. R., and Dillmann, R. (2011). Distributed generalization of learned planning models in robot programming by demonstration. In *2011 IEEE/RSJ International Conference on Intelligent Robots and Systems*, pages 4633–4638. IEEE.

Jäkel, R., Schmidt-Rohr, S. R., Lösch, M., and Dillmann, R. (2010). Representation and constrained planning of manipulation strategies in the context of programming by demonstration. In *Robotics and Automation (ICRA), 2010 IEEE International Conference On*, pages 162–169. IEEE.

Jäkel, R., Schmidt-Rohr, S. R., Rühl, S. W., Kasper, A., Xue, Z., and Dillmann, R. (2012). Learning of Planning Models for Dexterous Manipulation Based on Human Demonstrations. *International Journal of Social Robotics*, 4(4):437–448.

Kaiser, P., Kanoulas, D., Grotz, M., Muratore, L., Rocchi, A., Hoffman, E. M., Tsagarakis, N. G., and Asfour, T. (2016). An affordance-based pilot interface for high-level control of humanoid robots in supervised autonomy. In *Humanoid Robots (Humanoids), 2016 IEEE-RAS 16th International Conference On*, pages 621–628. IEEE.

Kaiser, P., Lewis, M., Petrick, R. P., Asfour, T., and Steedman, M. (2014). Extracting common sense knowledge from text for robot planning. In *2014 IEEE International Conference on Robotics and Automation (ICRA)*, pages 3749–3756. IEEE.

Kalman, R. E. (1960). A new approach to linear filtering and prediction problems. *Journal of basic Engineering*, 82(1):35–45.

Kang, S. B. and Ikeuchi, K. (1995). Toward automatic robot instruction from perception-temporal segmentation of tasks from human hand motion. *IEEE Transactions on Robotics and Automation*, 11(5):670–681.

Kasper, A., Xue, Z., and Dillmann, R. (2012). The KIT object models web database: An object model database for object recognition, localization and manipulation in service robotics. In *The International Journal of Robotics Research*.

Keogh, E., Chu, S., Hart, D., and Pazzani, M. (2004). Segmenting time series: A survey and novel approach. *Data mining in time series databases*, 57:1–22.

Klotzbücher, M. and Bruyninckx, H. (2012). Coordinating robotic tasks and systems with rFSM statecharts. *JOSER: Journal of Software Engineering for Robotics*, 3(1):28–56.

Koenig, N. and Mataric, M. (2006). Behavior-based segmentation of demonstrated tasks. *Proceedings of the International Conference on Development and Learning*.

Kohlmorgen, J. and Lemm, S. (2002). A Dynamic HMM for On-line Segmentation of Sequential Data. In *Advances in Neural Information Processing Systems 14 (NIPS 2001*, pages 793–800. MIT Press.

Konecny, S., Stock, S., Pecora, F., and Saffiotti, A. (2014). Planning domain + execution semantics: A way towards robust execution? In *Proc. of Qualitative Representations for Robots, AAAI Spring Symposium*.

Kozlov, A. (2013). *Akquirierung von Umgebungswissen aus der Aktionsausführung für Humanoide Roboter*. Masterarbeit, KIT.

Krüger, N., Geib, C., Piater, J., Petrick, R., Steedman, M., Wörgötter, F., Ude, A., Asfour, T., Kraft, D., Omrčen, D., and others (2011). Object–Action Complexes: Grounded abstractions of sensory–motor processes. *Robotics and Autonomous Systems*, 59(10):740–757.

Kulic, D. and Nakamura, Y. (2010). Incremental learning of human behaviors using hierarchical hidden Markov models. In *Intelligent Robots and Systems (IROS), 2010 IEEE/RSJ International Conference On*, pages 4649–4655. IEEE.

Kulic, D., Takano, W., and Nakamura, Y. (2009). Online segmentation and clustering from continuous observation of whole body motions. *Robotics, IEEE Transactions on*, 25(5):1158–1166.

Kuniyoshi, Y., Inaba, M., and Inoue, H. (1994). Learning by watching: Extracting reusable task knowledge from visual observation of human performance. *IEEE Transactions on Robotics and Automation*, 10(6):799–822.

Lan, R. and Sun, H. (2015). Automated human motion segmentation via motion regularities. *Visual Computer*, 31(1):35–53.

Larsen, E., Gottschalk, S., Lin, M. C., and Manocha, D. (1999). Fast proximity queries with swept sphere volumes. Technical report, Technical Report TR99-018, Department of Computer Science, University of North Carolina.

Lee, S., Suh, I., Calinon, S., and Johansson, R. (2012). Learning basis skills by autonomous segmentation of humanoid motion trajectories. In *IEEE-RAS International Conference on Humanoid Robots*, pages 112–119.

Lee, S. H., Suh, I. H., Calinon, S., and Johansson, R. (2015). Autonomous framework for segmenting robot trajectories of manipulation task. *Autonomous Robots*, 38(2):107–141.

Lehmann, A., Mikut, R., and Asfour, T. (2006). Petri Nets for Task Supervision in Humanoid Robots. In *Proc. of 37th International Symposium of Robotics (ISR)*, pages 183–202.

Li, Z., Wei, Z., Jia, W., and Sun, M. (2013). Daily life event segmentation for lifestyle evaluation based on multi-sensor data recorded by a wearable device. In *Proceedings of the Annual International Conference of the IEEE Engineering in Medicine and Biology Society, EMBS*, pages 2858–2861.

Lieberman, J. and Breazeal, C. (2004). Improvements on action parsing and action interpolation for learning through demonstration. In *4th IEEE/RAS International Conference on Humanoid Robots*, volume 1, pages 342–365. IEEE.

Lin, J., Bonnet, V., Panchea, A. M., Ramdani, N., Venture, G., and Kulic, D. (2016a). Human Motion Segmentation using Cost Weights Recovered from Inverse Optimal Control. In *IEEE-RAS 16th International Conference on Humanoid Robots (Humanoids)*.

Lin, J. and Kulić, D. (2012). Human pose recovery using wireless inertial measurement units. *Physiological Measurement*, 33(12):2099–2115.

Lin, J. F.-S., Joukov, V., and Kulić, D. (2014). Human motion segmentation by data point classification. In *IEEE Engineering in Medicine and Biology Conference*, pages 9–13. IEEE.

Lin, J. F.-S., Karg, M., and Kulić, D. (2016b). Movement Primitive Segmentation for Human Motion Modeling: A Framework for Analysis. *IEEE Transactions on Human-Machine Systems*, 46(3):325–339.

Lin, J. F.-S. and Kulić, D. (2014). Online segmentation of human motion for automated rehabilitation exercise analysis. *IEEE Transactions on Neural Systems and Rehabilitation Engineering*, 22(1):168–180.

Lisca, G., Nyga, D., Bálint-Benczédi, F., Langer, H., and Beetz, M. (2015). Towards robots conducting chemical experiments. In *Intelligent Robots and Systems (IROS), 2015 IEEE/RSJ International Conference On*, pages 5202–5208. IEEE.

Liu, R. and Zhang, X. (2016). Generating machine-executable plans from end-user's natural-language instructions. *arXiv preprint arXiv:1611.06468*.

Livingston, M. A., Sebastian, J., Ai, Z., and Decker, J. W. (2012). Performance measurements for the Microsoft Kinect skeleton. In *2012 IEEE Virtual Reality Workshops (VRW)*, pages 119–120. IEEE.

Loetzsch, M., Risler, M., and Jungel, M. (2006). XABSL-a pragmatic approach to behavior engineering. In *2006 IEEE/RSJ International Conference on Intelligent Robots and Systems*, pages 5124–5129. IEEE.

Loh, W.-Y. (2011). Classification and regression trees. *Wiley Interdisciplinary Reviews: Data Mining and Knowledge Discovery*, 1(1):14–23.

Lötzsch, M., Bach, J., Burkhard, H.-D., and Jüngel, M. (2003). Designing agent behavior with the extensible agent behavior specification language XABSL. In *Robot Soccer World Cup*, pages 114–124. Springer.

Luinge, H. and Veltink, P. (2005). Measuring orientation of human body segments using miniature gyroscopes and accelerometers. *Medical and Biological Engineering and Computing*, 43(2):273–282.

Lv, F. and Nevatia, R. (2006). Recognition and segmentation of 3-d human action using hmm and multi-class adaboost. In *European Conference on Computer Vision*, pages 359–372. Springer.

MacQueen, J. (1967). Some methods for classification and analysis of multivariate observations. In *Proceedings of the Fifth Berkeley Symposium on Mathematical Statistics and Probability*, volume 1, pages 281–297. Oakland, CA, USA.

Maji, S., b Bourdev, L., and Malik, J. (2011). Action recognition from a distributed representation of pose and appearance. In *Proceedings of the IEEE Computer Society Conference on Computer Vision and Pattern Recognition*, pages 3177–3184.

Mandery, C., Borràs, J., Jöchner, M., and Asfour, T. (2016a). Using Language Models to Generate Whole-Body Multi-Contact Motions. In *IEEE/RSJ*

International Conference on Intelligent Robots and Systems (IROS), pages 5411–5418.

Mandery, C., Terlemez, Ö., Do, M., Vahrenkamp, N., and Asfour, T. (2016b). Unifying Representations and Large-Scale Whole-Body Motion Databases for Studying Human Motion. *IEEE Transactions on Robotics*, 32(4):796–809.

Marszalek, M., Laptev, I., and Schmid, C. (2009). Actions in context. In *2009 IEEE Computer Society Conference on Computer Vision and Pattern Recognition Workshops, CVPR Workshops 2009*, pages 2929–2936.

MathWorks (2015a). MATLAB [software].

MathWorks (2015b). Simulink [software].

MathWorks (2015c). Stateflow [software].

Matsuo, K., Murakami, K., Hasegawa, T., Tahara, K., and Kurazume, R. (2009). Segmentation method of human manipulation task based on measurement of force imposed by a human hand on a grasped object. In *2009 IEEE/RSJ International Conference on Intelligent Robots and Systems*, pages 1767–1772. IEEE.

Matuszek, C., Herbst, E., Zettlemoyer, L., and Fox, D. (2013). Learning to parse natural language commands to a robot control system. In *Experimental Robotics*, pages 403–415. Springer.

Meier, F., Theodorou, E., Stulp, F., and Schaal, S. (2011). Movement segmentation using a primitive library. In *IEEE International Conference on Intelligent Robots and Systems*, pages 3407–3412.

Merz, T., Rudol, P., and Wzorek, M. (2006). Control System Framework for Autonomous Robots Based on Extended State Machines. In *ICAS*, page 14. IEEE Computer Society.

Metta, G., Fitzpatrick, P., and Natale, L. (2006). YARP: Yet Another Robot Platform. *International Journal on Advanced Robotics Systems.*

Microsoft (2012a). Robotics Developer Studio [software].

Microsoft (2012b). Visual Programming Language [software].

Mohseni-Kabir, A., Chernova, S., and Rich, C. (2014). Collaborative learning of hierarchical task networks from demonstration and instruction. In *Workshop on Human-Robot Collaboration for Industrial Manufacturing*, Berkeley.

Mohseni-Kabir, A., Rich, C., Chernova, S., Sidner, C. L., and Miller, D. (2015). Interactive hierarchical task learning from a single demonstration. In *Proceedings of the Tenth Annual ACM/IEEE International Conference on Human-Robot Interaction*, pages 205–212. ACM.

Motion Analysis Corporation (2016). Raptor. `http://www.motionanalysis.com/`.

Mourao, K. M. T. (2012). *Learning Action Representations Using Kernel Perceptrons*. PhD thesis, University of Edinburgh, Edinburgh.

Mueen, A., Keogh, E., Zhu, Q., Cash, S., and Westover, B. (2009). Exact discovery of time series motifs. *Proc. SIAM Conf. Data Mining*, pages 473–484.

Mustafa, W., Pugeault, N., Buch, A. G., and Krüger, N. (2015). Multi-view object instance recognition in an industrial context. *Robotica*, pages 1–22.

Mustafa, W., Wächter, M., Szedmak, Sandor, Agostini, Alejandro, Kraft, Dirk, Asfour, Tamim, Piater, Justus, Wörgötter, Florentin, and Krüger, Norbert (2016). Affordance Estimation For Vision-Based Object Replacement on a Humanoid Robot. In *47th International Symposium on Robotics (ISR)*, pages 1–9, Munich, Germany.

Naylor, A., Shao, L., Volz, R., Jungclas, R., Bixel, P., and Lloyd, K. (1987). PROGRESS–A graphical robot programming system. In *1987 IEEE International Conference on Robotics and Automation Proceedings*, volume 4, pages 1282–1291.

Newell, K. M., Slifkin, A. B., and Piek, J. P. (1998). *Motor Behavior and Human Skill: A Multidisciplinary Approach.* Champaign, IL: Human Kinetics.

Nicolescu, M. N. and Matarić, M. J. (2002). A hierarchical architecture for behavior-based robots. In *Proceedings of the First International Joint Conference on Autonomous Agents and Multiagent Systems: Part 1*, pages 227–233. ACM.

Nicolescu, M. N. and Mataric, M. J. (2003). Natural methods for robot task learning: Instructive demonstrations, generalization and practice. In *Proceedings of the Second International Joint Conference on Autonomous Agents and Multiagent Systems*, pages 241–248. ACM.

Nilsson, N. J. (1973). *A Hierarchical Robot Planning and Execution System.* SRI International Menlo Park, CA.

Nyga, D. and Beetz, M. (2012a). Everything robots always wanted to know about housework (but were afraid to ask). In *Proc. of IROS*, pages 243–250.

Nyga, D. and Beetz, M. (2012b). Everything robots always wanted to know about housework (but were afraid to ask). In *Intelligent Robots and Systems (IROS), 2012 IEEE/RSJ International Conference On*, pages 243–250. IEEE.

Nyga, D. and Beetz, M. (2015). Cloud-based Probabilistic Knowledge Services for Instruction Interpretation. *ISRR, Genoa, Italy.*

Object Management Group (OMG) (2015). OMG Unified Modeling Language Version 2.5.

OptiTrack (2016). Motion capture and real-time tracking with body, face, and object tracking systems. `https://www.optitrack.com/`.

Ormoneit, D., Sidenbladh, H., Black, M., and Hastie, T. (2001). Learning and tracking cyclic human motion. *Proc. Adv. Neural Inf. Process. Syst.*, pages 894–900.

Orr, M. J. (1996). *Introduction to Radial Basis Function Networks*. Technical Report, Center for Cognitive Science, University of Edinburgh.

Ovchinnikova, E., Wächter, M., Wittenbeck, V., and Asfour, T. (2015). Multi-Purpose Natural Language Understanding Linked to Sensorimotor Experience in Humanoid Robots. In *IEEE/RAS International Conference on Humanoid Robots (Humanoids)*, pages 365–372.

Paikan, A. (2014). *Enhancing Software Module Reusability and Development in Robotic Applications*. dissertation, Istituto Italiano di Tecnologia.

Paikan, A., Fitzpatrick, P., Metta, G., and Natale, L. (2014a). Data flow port monitoring and arbitration. *Software Engineering for Robotics*, 5(1):80–88.

Paikan, A., Metta, G., and Natale, L. (2014b). A Representation Of Robotic Behaviors Using Component Port Arbitration. In *5th International Workshop on Domain-Specific Languages and Models for ROBotic Systems (DSLRob)*.

Pardowitz, M., Haschke, R., Steil, J., and Ritter, H. (2008). Gestalt-based action segmentation for robot task learning. In *Humanoids 2008-8th IEEE-RAS International Conference on Humanoid Robots*, pages 347–352. IEEE.

Pedersen, T., Patwardhan, S., and Michelizzi, J. (2004). WordNet:: Similarity: Measuring the relatedness of concepts. In *Demonstration Papers at HLT-NAACL 2004*, pages 38–41. Association for Computational Linguistics.

Pednault, E. P. (1987). Formulating multiagent, dynamic-world problems in the classical planning framework. *Reasoning about actions and plans*, pages 47–82.

Pei, M., Si, Z., Yao, B. Z., and Zhu, S.-C. (2013). Learning and parsing video events with goal and intent prediction. *Computer Vision and Image Understanding*, 117(10):1369–1383.

Petrick, R. P. and Bacchus, F. (2002). A Knowledge-Based Approach to Planning with Incomplete Information and Sensing. In *Proc. of AIPS*, pages 212–222.

Pomplun, M. and Mataric, M. J. (2000). Evaluation metrics and results of human arm movement imitation. In *Proceedings, First IEEE-RAS International Conference on Humanoid Robotics (Humanoids*.

Pot, E., Monceaux, J., Gelin, R., and Maisonnier, B. (2009). Choregraphe: A graphical tool for humanoid robot programming. In *Robot and Human Interactive Communication, 2009. RO-MAN 2009. The 18th IEEE International Symposium On*, pages 46–51. IEEE.

Quantum Leaps (2015). QM statemachines [software].

Quigley, M., Conley, K., Gerkey, B. P., Faust, J., Foote, T., Leibs, J., Wheeler, R., and Ng, A. Y. (2009). ROS: An open-source Robot Operating System. In *ICRA Workshop on Open Source Software*, volume 3, page 5.

Rabiner, L. (1989). A Tutorial on Hidden Markov Models and Selected Applications in Speech Recognition. *Proceedings of the IEEE*, 77(2):257–286.

Rahul, R., Whitchurch, A., and Rao, M. (2014). An open source graphical robot programming environment in introductory programming curriculum for undergraduates. In *MOOC, Innovation and Technology in Education (MITE), 2014 IEEE International Conference On*, pages 96–100.

Ramirez-Amaro, K., Beetz, M., and Cheng, G. (2014). Automatic segmentation and recognition of human activities from observation based on semantic reasoning. In *2014 IEEE/RSJ International Conference on Intelligent Robots and Systems*, pages 5043–5048. IEEE.

Ramirez-Amaro, K., Beetz, M., and Cheng, G. (2015). Transferring skills to humanoid robots by extracting semantic representations from observations of human activities. *Artificial Intelligence*.

Raptis, M. and Sigal, L. (2013). Poselet key-framing: A model for human activity recognition. In *Proceedings of the IEEE Computer Society Conference on Computer Vision and Pattern Recognition*, pages 2650–2657.

Ratanamahatana, C. and Keogh, E. (2004). Making time-series classification more accurate using learned constraints. In *SIAM Proceedings Series*, pages 11–22.

Ricci, L., Formica, D., Tamilia, E., Taffoni, F., Sparaci, L., Capirci, O., and Guglielmelli, E. (2013). An experimental protocol for the definition of upper limb anatomical frames on children using magneto-inertial sensors. In *Engineering in Medicine and Biology Society (EMBC), 2013 35th Annual International Conference of the IEEE*, pages 4903–4906. IEEE.

Roetenberg, D., Luinge, H., and Slycke, P. (2009). Xsens MVN: Full 6DOF human motion tracking using miniature inertial sensors. *Xsens Motion Technologies BV, Tech. Rep.*

Russell, S. J., Norvig, P., Canny, J. F., Malik, J. M., and Edwards, D. D. (2003). *Artificial Intelligence: A Modern Approach*, volume 2. Prentice hall Upper Saddle River.

Samek, M. (2002). *Practical Statecharts in C/C++: Quantum Programming for Embedded Systems*. CRC Press.

Schlichting, N., Damsma, A., Aksoy, E. E., Wächter, M., Asfour, T., and van Rijn, H. (2018). Temporal context influences the perceived duration of everyday actions: Assessing the ecological validity of lab-based timing phenomena. *Journal of Cognition*, 1(1).

Scholl, K.-U., Albiez, J., and Gassmann, B. (2001). Mca-an expandable modular controller architecture. In *3rd Real-Time Linux Workshop*.

Sell, J. and O'Connor, P. (2014). The Xbox One System on a Chip and Kinect Sensor. *IEEE Micro*, 34(2):44–53.

Simmons, R. and Apfelbaum, D. (1998). A task description language for robot control. In *Intelligent Robots and Systems, 1998. Proceedings., 1998 IEEE/RSJ International Conference On*, volume 3, pages 1931–1937. IEEE.

Stampfer, D. and Schlegel, C. (2014). Dynamic State Charts: Composition and coordination of complex robot behavior and reuse of action plots. *Intelligent Service Robotics*, 7(2):53–65.

Summers-Stay, D., Teo, C. L., Yang, Y., Fermüller, C., and Aloimonos, Y. (2012). Using a minimal action grammar for activity understanding in the real world. In *2012 IEEE/RSJ International Conference on Intelligent Robots and Systems*, pages 4104–4111. IEEE.

Szedmak, S., Ugur, E., and Piater, J. (2014). Knowledge propagation and relation learning for predicting action effects. In *Intelligent Robots and Systems (IROS 2014), 2014 IEEE/RSJ International Conference On*, pages 623–629. IEEE.

Terlemez, Ö., Ulbrich, S., Mandery, C., Do, M., Vahrenkamp, N., and Asfour, T. (2014). Master Motor Map (MMM)—Framework and toolkit for capturing, representing, and reproducing human motion on humanoid robots. In *2014 IEEE-RAS International Conference on Humanoid Robots*, pages 894–901. IEEE.

Thomas, U., Hirzinger, G., Rumpe, B., Schulze, C., and Wortmann, A. (2013). A new skill based robot programming language using uml/p statecharts. In *Robotics and Automation (ICRA), 2013 IEEE International Conference On*, pages 461–466. IEEE.

Tipping, M. E. and Bishop, C. M. (1999). Probabilistic principal component analysis. *Journal of the Royal Statistical Society: Series B (Statistical Methodology)*, 61(3):611–622.

Ulbrich, S., Ruiz, V., Asfour, T., Torras, C., and Dillmann, R. (2012). Kinematic Bézier Maps. *IEEE Transactions on Systems, Man, and Cybernetics*, 42(4):1215–1230.

Vahrenkamp, N., Kröhnert, M., Ulbrich, S., Asfour, T., Metta, G., Dillmann, R., and Sandini, G. (2012). Simox: A Robotics Toolbox for Simulation, Motion and Grasp Planning. In *International Conference on Intelligent Autonomous Systems (IAS)*, pages 585–594.

Vahrenkamp, N., Wächter, M., Kröhnert, M., Kaiser, P., Welke, K., and Asfour, T. (2014). High-Level Robot Control with ArmarX. In *INFORMATIK – Workshop on Robot Control Architectures*, pages 1–12.

Vahrenkamp, N., Wächter, M., Kröhnert, M., Welke, K., and Asfour, T. (2015). The ArmarX Framework - Supporting high level robot programming through state disclosure. *Information Technology*, 57(2):99–111.

van Rijsbergen, C. (1979). Information retrieval. *The Information Retrieval Group*.

Vicente, I., Kyrki, V., Kragic, D., and Larsson, M. (2007). Action recognition and understanding through motor primitives. *Advanced Robotics*, 21(15):1687–1707.

VICON (2016). Motion Capture Systems. `http://www.vicon.com`.

Vögele, A., Krüger, B., and Klein, R. (2014). Efficient unsupervised temporal segmentation of human motion. *ACM SIGGRAPH Symp. Comput. Animation*.

Von der Beeck, M. (1994). A comparison of statecharts variants. In *Formal Techniques in Real-Time and Fault-Tolerant Systems*, pages 128–148. Springer.

Wächter, M. and Asfour, T. (2015). Hierarchical segmentation of manipulation actions based on object relations and motion characteristics. In *2015 International Conference on Advanced Robotics (ICAR)*, pages 549–556.

Wächter, M., Ottenhaus, S., Kröhnert, M., Vahrenkamp, N., and Asfour, T. (2016). The ArmarX Statechart Concept: Graphical Programing of Robot Behavior. *Frontiers in Robotics and AI*, 3:33.

Wächter, M., Ovchinnikova, E., Wittenbeck, V., Kaiser, P., Szedmak, S., Mustafa, W., Kraft, D., Krüger, N., Piater, J., and Asfour, T. (2018). Integrating multi-purpose natural language understanding, robot's memory, and symbolic planning for task execution in humanoid robots. *Robotics and Autonomous Systems*, 99:148–165.

Wächter, M., Schulz, S., Asfour, T., Aksoy, E., Wörgötter, F., and Dillmann, R. (2013). Action sequence reproduction based on automatic segmentation and Object-Action Complexes. In *2013 13th IEEE-RAS International Conference on Humanoid Robots (Humanoids)*, pages 189–195.

Wang, J., Nie, X., Xia, Y., and Wu, Y. (2014). Mining discriminative 3D Poselet for cross-view action recognition. In *2014 IEEE Winter Conference on Applications of Computer Vision, WACV 2014*, pages 634–639.

Welke, K., Kaiser, P., Kozlov, A., Adermann, N., Asfour, T., Lewis, M., and Steedman, M. (2013a). Grounded Spatial Symbols for Task Planning Based on Experience. In *IEEE/RAS International Conference on Humanoid Robots (Humanoids)*, pages 484–491.

Welke, K., Schiebener, D., Asfour, T., and Dillmann, R. (2013b). Gaze selection during manipulation tasks. In *IEEE International Conference on Robotics and Automation (ICRA)*, pages 652–659.

Welke, K., Vahrenkamp, N., Wächter, M., Kröhnert, M., and Asfour, T. (2013c). The ArmarX Framework-Supporting high level robot programming through state disclosure. In *GI-Jahrestagung*, pages 2823–2837.

Winter, D. A. (1991). *Biomechanics and Motor Control of Human Gait: Normal, Elderly and Pathological.* Waterloo Biomechanics, Waterloo , Ontario.

Wörgötter, F., Agostini, A., Krüger, N., Shylo, N., and Porr, B. (2009). Cognitive agents—a procedural perspective relying on the predictability of Object-Action-Complexes (OACs). *Robotics and Autonomous Systems*, 57(4):420–432.

Wörgötter, F., Aksoy, E., Krüger, N., Piater, J., Ude, A., and Tamosiunaite, M. (2013). A simple ontology of manipulation actions based on hand-object relations. *IEEE Transactions on Autonomous Mental Development*, 5(2):117–134.

Wörgötter, F., Geib, C., Tamosiunaite, M., Aksoy, E. E., Piater, J., Xiong, H., Ude, A., Nemec, B., Kraft, D., Krüger, N., Wächter, M., and Asfour, T. (2015). Structural bootstrapping - a novel concept for the fast acquisition of action-knowledge. *IEEE Trans. on Autonomous Mental Development*, 7(2):140–154.

World Wide Web Consortium (W3) (2015). State Chart XML (SCXML): State Machine Notation for Control Abstraction.

Xiong, H., Szedmak, S., and Piater, J. (2013). Homogeneity Analysis for Object-action Relation Reasoning in Kitchen Scenarios. In *Proceedings of the 2Nd Workshop on Machine Learning for Interactive Systems: Bridging the Gap Between Perception, Action and Communication*, MLIS '13, pages 37–44, New York, NY, USA. ACM.

Xperience (2011). *The Xperience Project.*

Yakindu (2015). Yakindu Statechart Editor Tools.

Yamamoto, M., Mitomi, H., Fujiwara, F., and Sato, T. (2006). Bayesian classification of task-oriented actions based on stochastic context-free grammar. In *7th International Conference on Automatic Face and Gesture Recognition, 2006. FGR 2006*, pages 317–322.

Yang, W., Wang, Y., and Mori, G. (2010). Recognizing human actions from still images with latent poses. In *Proceedings of the IEEE Computer Society Conference on Computer Vision and Pattern Recognition*, pages 2030–2037.

Yi, B.-K., Jagadish, H. V., and Faloutsos, C. (1998). Efficient retrieval of similar time sequences under time warping. In *Data Engineering, 1998. Proceedings., 14th International Conference On*, pages 201–208. IEEE.

Yuwono, M., Su, S., Moulton, B., and Nguyen, H. (2013). Unsupervised segmentation of heel-strike IMU data using rapid cluster estimation of wavelet features. In *Proceedings of the Annual International Conference of the IEEE Engineering in Medicine and Biology Society, EMBS*, pages 953–956.

Zhang, Y., Qu, W., and Wang, D. (2014). Action-scene model for human action recognition from videos. *Second AASRI Conference on Computational Intelligence and Bioinformatics*, 6:111–117.

Zhang, Z. (2012). Microsoft kinect sensor and its effect. *IEEE multimedia*, 19(2):4–10.

Zhou, F., De la Torre, F., and Hodgins, J. K. (2013). Hierarchical aligned cluster analysis for temporal clustering of human motion. *IEEE Transactions on Pattern Analysis and Machine Intelligence*, 35(3):582–596.

Ziaeefard, M. and Bergevin, R. (2015). Semantic human activity recognition: A literature review. *Pattern Recognition*, 48(8):2329–2345.

Ziparo, V. A. and Iocchi, L. (2006). Petri net plans. In *Proceedings of Fourth International Workshop on Modelling of Objects, Components, and Agents (MOCA)*, pages 267–290.

Zöllner, R., Pardowitz, M., Knoop, S., and Dillmann, R. (2005). Towards cognitive robots: Building hierarchical task representations of manipulations from human demonstration. In *Proceedings of the 2005 IEEE International Conference On Robotics and Automation*, pages 1535–1540. IEEE.